AF360838

Carnot Cycle and Heat Engine Fundamentals and Applications II

Carnot Cycle and Heat Engine Fundamentals and Applications II

Editor

Michel Feidt

MDPI • Basel • Beijing • Wuhan • Barcelona • Belgrade • Manchester • Tokyo • Cluj • Tianjin

Editor
Michel Feidt
University of Lorraine
France

Editorial Office
MDPI
St. Alban-Anlage 66
4052 Basel, Switzerland

This is a reprint of articles from the Special Issue published online in the open access journal *Entropy* (ISSN 1099-4300) (available at: https://www.mdpi.com/journal/entropy/special_issues/Carnot_Cycle_II).

For citation purposes, cite each article independently as indicated on the article page online and as indicated below:

LastName, A.A.; LastName, B.B.; LastName, C.C. Article Title. *Journal Name* **Year**, *Volume Number*, Page Range.

ISBN 978-3-0365-3260-8 (Hbk)
ISBN 978-3-0365-3261-5 (PDF)

Contents

About the Editor

Michel Feidt is an emeritus professor at the University of Lorraine, France, where he has spent his entire career in education and research. His main interests are thermodynamics and energy. He is a specialist in infinite physical dimensions optimal thermodynamics (FDOT), which he considers from a fundamental point of view, illustrating the necessity of considering irreversibility to optimize systems and processes and to characterize upper bound efficiencies. He has published many articles in journals and books: more than 120 papers and more than 5 books. He actively participates in numerous international and national conferences on the same subject. He has developed 55 final contract reports and has advised 43 theses. He has been a member of more than 110 doctoral committees. He is a member of the scientific committee of more than five scientific journals and editor-in-chief of one journal.

Editorial

The Carnot Cycle and Heat Engine Fundamentals and Applications II

Michel Feidt

Laboratory of Energetics, Theoretical and Applied Mechanics (LEMTA), URA CNRS 7563, University of Lorraine, 54518 Vandoeuvre-lès-Nancy, France; michel.feidt@univ-lorraine.fr

Citation: Feidt, M. The Carnot Cycle and Heat Engine Fundamentals and Applications II. *Entropy* **2022**, *24*, 230. https://doi.org/10.3390/e24020230

Received: 27 January 2022
Accepted: 28 January 2022
Published: 2 February 2022

Publisher's Note: MDPI stays neutral with regard to jurisdictional claims in published maps and institutional affiliations.

This editorial introduces the second Special Issue entitled "Carnot Cycle and Heat Engine Fundamentals and Applications II" https://www.mdpi.com/si/entropy/Carnot_Cycle_II (accessed on 29 January 2022).

The editorial of this Special Issue comes after the review process. Nine papers have been published between 26 February 2021 and 4 January 2022 due to the COVID-19 pandemic. These papers are listed hereafter in the inverse order of date of publication. Thanks to all the authors for the various viewpoints expressed that unveil fundamental and application aspects of the Carnot cycle and heat engines.

Authors are from Europe (four papers) and China (five papers). Each paper has been viewed by 400 to 1100 persons, except the last published one. Four papers have been presently cited 8 to 20 times.

Five papers address heat engines and Carnot configurations [1–5]. Papers by [2,4,5] concern, respectively, diesel engine, Lenoir, and Brayton cycles. The papers by [1,2] are related to Carnot engines. However, these five papers address real, irreversible cases. Three papers from Chinese authors [2,4,5] deal with finite-time thermodynamics (FTT). Papers by [2,5] use numerical methods such as genetic algorithm NASCA II (through LINMAP method, TOPSIS method, and Shannon entropy method) to optimize engines. The various objectives considered include power, power density, ecological function, and first law efficiency.

The paper that discusses the Lenoir cycle is from a more conventional point of view. It deals with the steady flow (such as Chambadal's original modeling). Objectives are power and first law efficiency. The corresponding allocation of heat transfer conductance is proposed, due to finite size constraints (i.e., the Utotal imposed).

In [2], the authors consider the optimization of an irreversible Carnot engine, comparing the FTT approach to the finite speed thermodynamics approach (FST). The direct method combined with the first law efficiency takes irreversibility into account (heat transfer gradients, pressure losses, and mechanical frictions). The main results include the following:

-   Maximum energy efficiency differs from maximum power through different variable piston speed values;
-   Results obtained through the FST method are different from those obtained from the Curzon–Ahlborn model (with time duration), due to the steady-state hypothesis.

Paper [1] concerns the modified Chambadal model of Carnot engines. It, too, addresses irreversibility but from a global point of view. This paper completes and improves the one proposed in the preceding Special Issue. A sequential optimization corresponding to various finite physical dimensions constraints is developed with the three objectives of energy, first law efficiency, and power. Two new concepts of entropic action are proposed and used—entropic action relative to production of entropy and entropic action relative to the transfer of entropy.

Papers by [6,7] extend the configuration from engines to reverse cycle machines including Stirling refrigerating machine [6] and Brayton refrigerating machine [7]. The paper by [7] combines, in fact, direct and inverse Brayton cycles, constituting more of a system, with regeneration purposes (regeneration before the inverse cycle). Constraints regarding pressure losses and size are considered.

The study by [6] is, in fact, related to the paper by [7], published in the preceding Special Issue: It discusses a finite physical dimension in a Stirling refrigerating machine according to Schmidt modeling. The paper uses entropy and exergy analysis. The most important irreversibility mechanisms are thermal ones and, more precisely, those due to regeneration.

Papers of [8,9] are specific but very interesting.

In [8], the authors discuss the chemical aspects of entropy and exergy analysis, including reconsideration of concepts and definitions relating the entropy–exergy relationship, with applications in industrial engineering and biotechnologies. The main objective is to evaluate the performance associated with all interactions between the system and the external environment. This is a crucial challenge today due to environmental concerns.

Paper by [9] is related to a very important and up-to-date subject—superconducting quantum circuits. It concerns a new approach mixing finite-time and quantum thermodynamics: quantum heat engine cycle. Closely linked to these fundamental aspects are corresponding applications for quantum computers.

To conclude, this second Special Issue confirms and improves the preceding one in terms of the following aspects:

- Systematic consideration of irreversibility (more than endo-reversibility);
- Two ways of optimization—namely, sequential (mainly analytical) and multiobjective (mainly numerical) approaches;
- Various objectives including energy, power, and first law efficiency for the most used approach;
- Various constraints; from a general point of view, the use of what we introduce as finite physical dimensions of optimal thermodynamics (FDOT) with finite constraints (see the book of the author of this editorial);
- The evolution of research from basic cycle to complex systems.

Perhaps these features could pave the way toward a third Special Issue, to expand and build upon concepts and approaches presented thus far.

Funding: This research received no external funding.

Acknowledgments: We express our thanks to the authors of the above contributions, and to the journal *Entropy* and MDPI for their support during this Special Issue.

Conflicts of Interest: The author declares no conflict of interest.

References

1. Feidt, M.; Costea, M. A New Step in the Optimization of the Chambadal Model of the Carnot Engine. *Entropy* **2022**, *24*, 84. [CrossRef] [PubMed]
2. Shi, S.; Chen, L.; Ge, Y.; Feng, H. Performance Optimizations with Single-, Bi-, Tri-, and Quadru-Objective for Irreversible Diesel Cycle. *Entropy* **2021**, *23*, 826. [CrossRef] [PubMed]
3. Costea, M.; Petrescu, S.; Feidt, M.; Dobre, C.; Borcila, B. Optimization Modeling of Irreversible Carnot Engine from the Perspective of Combining Finite Speed and Finite Time Analysis. *Entropy* **2021**, *23*, 504. [CrossRef] [PubMed]
4. Wang, R.; Ge, Y.; Chen, L.; Feng, H.; Wu, Z. Power and Thermal Efficiency Optimization of an Irreversible Steady-Flow Lenoir Cycle. *Entropy* **2021**, *23*, 425. [CrossRef] [PubMed]
5. Tang, C.; Chen, L.; Feng, H.; Ge, Y. Four-Objective Optimizations for an Improved Irreversible Closed Modified Simple Brayton Cycle. *Entropy* **2021**, *23*, 282. [CrossRef] [PubMed]
6. Dobre, C.; Grosu, L.; Dobrovicescu, A.; Chişiu, G.; Constantin, M. Stirling Refrigerating Machine Modeling Using Schmidt and Finite Physical Dimensions Thermodynamic Models: A Comparison with Experiments. *Entropy* **2021**, *23*, 368. [CrossRef] [PubMed]

7. Chen, L.; Feng, H.; Ge, Y. Power and Efficiency Optimization for Open Combined Regenerative Brayton and Inverse Brayton Cycles with Regeneration before the Inverse Cycle. *Entropy* **2020**, *22*, 677. [CrossRef] [PubMed]
8. Palazzo, P. Chemical and Mechanical Aspect of Entropy-Exergy Relationship. *Entropy* **2021**, *23*, 972. [CrossRef] [PubMed]
9. Chen, J.-F.; Li, Y.; Dong, H. Simulating Finite-Time Isothermal Processes with Superconducting Quantum Circuits. *Entropy* **2021**, *23*, 353. [CrossRef] [PubMed]

 *entropy*

MDPI

Article

A New Step in the Optimization of the Chambadal Model of the Carnot Engine

Michel Feidt [1] and Monica Costea [2,*]

[1] Laboratory of Energetics, Theoretical and Applied Mechanics (LEMTA), URA CNRS 7563, University of Lorraine, 54518 Vandoeuvre-lès-Nancy, France; michel.feidt@univ-lorraine.fr
[2] Department of Engineering Thermodynamics, University Politehnica of Bucharest, 060042 Bucharest, Romania
* Correspondence: monica.costea@upb.ro; Tel.: +40-021-402-9339

Abstract: This paper presents a new step in the optimization of the Chambadal model of the Carnot engine. It allows a sequential optimization of a model with internal irreversibilities. The optimization is performed successively with respect to various objectives (e.g., energy, efficiency, or power when introducing the duration of the cycle). New complementary results are reported, generalizing those recently published in the literature. In addition, the new concept of entropy production action is proposed. This concept induces new optimums concerning energy and power in the presence of internal irreversibilities inversely proportional to the cycle or transformation durations. This promising approach is related to applications but also to fundamental aspects.

Keywords: optimization; Carnot engine; Chambadal model; entropy production action; efficiency at maximum power

1. Introduction

Sadi Carnot had a crucial contribution to thermostatics that designated him as a co-founding researcher of equilibrium thermodynamics. He has shown that the efficiency of a thermo-mechanical engine is bounded by the Carnot efficiency η_C [1]. Assuming an isothermal source at T_{HS}, and an isothermal sink at $T_{CS} < T_{HS}$, and in between the cycle composed by two isothermals in perfect thermal contact with the source and sink, and two isentropics, he obtained:

$$\eta_C = 1 - \frac{T_{CS}}{T_{HS}}. \tag{1}$$

Since that time, many papers have used the keyword "Carnot engine" (1290 papers from Web of Science on 17 September 2021). That same day on Web of Science, we noted 104 papers related to the keyword "Carnot efficiency".

Among these papers, some are related to the connection between energy, efficiency, and power optimization. The most cited paper is probably that of Curzon and Ahlborn [2,3]. These authors proposed in 1975 an expression of the efficiency according to the first law of thermodynamics $\eta_I(MaxW)$ at the maximum mechanical energy and at the maximum power $\dot{W}$ for the endo-reversible configuration of the Carnot cycle (no internal irreversibility for the converter in contact with two isothermal heat reservoirs):

$$\eta_{I,endo}(MaxW) = 1 - \sqrt{\frac{T_{CS}}{T_{HS}}} \tag{2}$$

This result is well-known as the *nice radical*, and it has been recently reconsidered in the previous Special Issue *Carnot Cycle and Heat Engine Fundamentals and Applications*

Citation: Feidt, M.; Costea, M. A New Step in the Optimization of the Chambadal Model of the Carnot Engine. *Entropy* 2022, 24, 84. https://doi.org/10.3390/e24010084

Academic Editor: José Miguel Mateos Roco

Received: 21 November 2021
Accepted: 24 December 2021
Published: 4 January 2022

Publisher's Note: MDPI stays neutral with regard to jurisdictional claims in published maps and institutional affiliations.

I [3] and particularly in [4]. This last paper reports on the progress in Carnot and Chambadal modeling of thermomechanical engines by considering entropy production and heat transfer entropy in the adiabatic case (without heat losses).

The proposed paper gives back the basis of the modeling and a summary of the main results obtained recently for an endo-irreversible Carnot engine. Furthermore, the performance analysis of an extended Chambadal configuration is considered by including the converter irreversibilities. Emphasis is placed on the entropy production method, which is preferred over the ratio method.

2. Summary of Obtained Results for Carnot Endo-Irreversible Configuration

The consideration of *endo-irreversible* Carnot engine modeling was recently developed [5]. The approach considering as a reference the heat transfer entropy released at the sink ΔS_S (maximum entropy available at the source in the reversible case) [5] confirmed that the work per cycle results (see Appendix A):

$$W = (T_{HS}{-}T_{CS})(\Delta S_S{-}\Delta S_I),\tag{3}$$

where ΔS_I is the entropy production due to the internal irreversibilities of the cycle throughout the four thermodynamic transformations (two adiabatic and two isothermal ones).

For an engine without thermal losses, the following expression of the thermal efficiency was retrieved:

$$\eta_I = \eta_C(1 - d_I),\tag{4}$$

where $d_I = \frac{\Delta S_I}{\Delta S_S}$ is a coefficient of the converter's internal irreversibility during the cycle.

This parameter was introduced by Novikov [6] and Ibrahim et al. [7] in slightly different forms.

The reversible limit ($d_I = 0$) in Equation (4) restores the Carnot cycle efficiency associated with equilibrium thermodynamics.

Since the reversibility is unattainable, it appears that the optimization (maximization) of the mechanical energy at the given parameters (ΔS_S, T_{HS}, and T_{CS}) is related to the minimization of the entropy production (as was proposed by Gouy [8]).

The assumption that each of the four transformations of the endo-irreversible cycle takes place with a duration τ_i ($i = 1$–4), leading to the inverse proportionality to τ_i of the corresponding entropy production:

$$\Delta S_{Ii} = \frac{C_{Ii}}{\tau_i},\tag{5}$$

where C_{Ii} represents the irreversibility coefficients, whose unit is Js/K [5].

These coefficients are *irreversible entropic actions* by analogy to the energy (mechanical) action (Js).

By performing cycle energy optimization using the Lagrange multipliers method with the constraint of the cycle's finite time duration τ, one obtains the maximum work per cycle [5]:

$$Max_1 W = W_{endo} - \frac{\Delta T_S}{\tau}\left(\sum_i \sqrt{C_{Ii}}\right)^2,\tag{6}$$

where $\Delta T_S = T_{HS} - T_{CS}$.

The efficiency at the maximum finite time work becomes

$$\eta_I(Max_1 W) = \eta_C\left(1 - \frac{\left(\sum_i \sqrt{C_{Ii}}\right)^2}{\tau \cdot \Delta S_S}\right),\tag{7}$$

where $\tau \Delta S_S$ is the available entropic transfer action of the cycle.

The new result provided by Equation (7) gives back the Carnot efficiency limit for the reversible case ($C_{Ii} = 0$). These calculations have been pursued for the case of power optimization, where ΔS_S, T_{HS}, and T_{CS} remain parameters. It was shown that a value of the cycle duration τ^* corresponding to $Max\overline{W}$, the mean power output over the cycle, exists, and it is expressed as

$$\tau^* = 8\frac{C_{Ii}}{\Delta S_S},\tag{8}$$

and

$$Max\overline{W} = \frac{\Delta T_S \cdot \Delta S_S^2}{16\,C_{Ii}}.\tag{9}$$

Equation (9) proves that $Max\overline{W}$ is a decreasing function of the total entropic action of the cycle and that the associated efficiency with the maximum of the mean power corresponds to half the Carnot efficiency, as appeared repeatedly in some recent works [9–11].

3. Summary of the Obtained Results for the Chambadal Configuration

In the present paper, we intend to reconsider the approach of the Chambadal model of a Carnot engine [12]. This configuration is common for thermomechanical engines, since the cold sink mainly refers to the environment (i.e., the atmosphere or water sink). This corresponds to the Chambadal approach (Figure 1), with a temperature gradient at the hot source (T_{HS}, T_H) but with perfect thermal contact at the sink (T_{CS} or T_0 at ambient temperature).

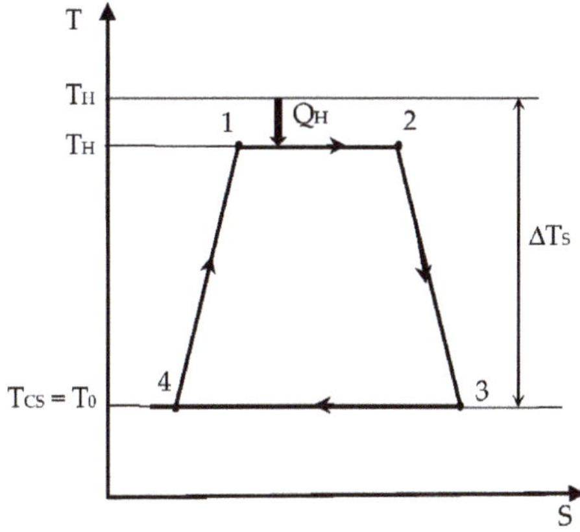

Figure 1. Representation of the associated cycle with the Chambadal engine in a *T-S* diagram.

We propose here to extend the results (Equations (6)–(9)) to enhance the Chambadal configuration modeling. This extension starts from the endo-irreversible case, to which external irreversibilities due to heat transfer between the hot finite source and the converter are added. Thus, the new results obtained complete the endo-irreversible Carnot model [5] and an earlier paper on Chambadal configuration [12].

3.1. The Modified Chambadal Engine

To help understand the extension of the modeling in Section 3, we report here the case with the following hypothesis:

1. *Adiabaticity* (no thermal losses);
2. *Linear heat transfer law* at the source such that

$$Q_H = G_H(T_{HS}-T_H),\tag{10}$$

where G_H is the heat transfer conductance expressed by $G_H = K_H\tau$ when we consider the mean value over the cycle duration τ or $G_H = K'_H\tau_H$ when we consider the mean value over the isothermal heat transfer at the hot source (as was performed by Curzon and Ahlborn [2]).

Equation (10) corresponds to the heat expense of the engine.

Note that other heat transfer laws, namely the Stefan–Boltzmann radiation law, the Dulong–Petit law, and another phenomenological heat transfer law can be considered in the maximum power regime search [13];

3. *Presence of irreversibility* in the converter (internal irreversibility).

Two approaches are proposed in the literature, which introduce the internal irreversibility of the engine by (1) the *irreversibility ratio* I_H, [6,7], respectively (2) the *entropy production over the cycle* ΔS_I, [5].

We preconized this second approach for a long time. We also note that the original model of Chambadal is endo-reversible [14]. Hence, we prefer to name the present model the "modified Chambadal model" due to some other differences that will be specified hereafter.

Note that only the second approach regarding the presence of irreversibilities in the converter will be considered in the following section.

3.2. Optimization of the Work per Cycle of the Modified Chambadal Engine with the Entropy Production Method

The first law of thermodynamics applied to the cycle implies conservation of energy, written as

$$W = Q_{conv} - Q_S \tag{11}$$

where Q_{conv} and Q_S are defined in Appendix A.

One supposes here that ΔS_I is a parameter representing the total production of entropy over the cycle composed by four irreversible transformations. Thus, the entropy balance corresponds to

$$\frac{Q_{conv}}{T_H} + \Delta S_I = \frac{Q_S}{T_0}. \tag{12}$$

By combining Equations (11) and (12), we easily obtained

$$W = Q_{conv}\left(1 - \frac{T_0}{T_H}\right) - T_0\Delta S_I. \tag{13}$$

If Q_{conv} (ΔS_{conv}) is a given parameter, we retrieve the Gouy-Stodola theorem stating that *Max W* corresponds to min ΔS_I with the known consequences reported in Section 4.1.

3.3. Optimization of the Work per Cycle of the Modified Chambadal Engine with the Heat Transfer Constraint

In this case, the energy balance between the source and isothermal transformation implies the combination of Equations (13) and (A1):

$$W = (Q_H - T_H\Delta S_{IH})\left(1 - \frac{T_0}{T_H}\right) - T_0\Delta S_I. \tag{14}$$

Knowing Q_H from Equation (10), one obtains

$$W = G_H(T_{HS} - T_H)\left(1 - \frac{T_0}{T_H}\right) - T_H\Delta S_{IH} - T_0\Delta S_I', \tag{15}$$

where $\Delta S_I' = \Delta S_{IEx} + \Delta S_{IC} + \Delta S_{ICo}$.

The maximum of W with respect to T_H is obtained for

$$T_H^* = \sqrt{\frac{T_{HS}T_0}{1 + s_I}}, \tag{16}$$

where $s_I = \frac{\Delta S_{IH}}{G_H}$, a specific ratio relative to the irreversible isothermal transformation T_H.

Finally, the expression of $Max_1 W$ yields

$$Max_1 W = G_H \left(\sqrt{T_{HS}} - \sqrt{(1 + s_I)T_0} \right)^2 - T_0 \Delta S_I. \tag{17}$$

4. Complement to the Previous Results

Now, we will consider the time variable related to entropy production for each thermodynamic transformation, defined as $\Delta S_{Ii} = \frac{C_{Ii}}{\tau_i}$. This form of the entropy production satisfying the second law induces that the entropy production method is well adapted to subsequent optimizations of energy and power as well.

4.1. Work Optimization Relative to the Time Variables

The expression of $Max_1 W$ with G_H as an extensive parameter (Equation (17)) shows that $Max_1 W$ is always the optimum in the endo-reversible case. Nevertheless, if there are separate irreversibilities for each cycle transformation (as is the case with finite entropic actions), the irreversibility on the high temperature isotherm possesses a specific role (see Equation (17) and the s_I ratio).

The constraint on the transformation duration or preferably frequencies f_i (finite cycle duration) allows one to seek for the optimal transformation duration allocation (see Appendix B for the derivation).

We obtained $Max_2 W$ for the following optimal durations:

$$\tau_H{}^* = \sqrt{\sqrt{T_0 T_{HS}} \frac{C_{IH}}{\lambda}}, \tag{18}$$

and

$$\tau_i{}^* = \sqrt{\frac{T_0 C_{Ii}}{\lambda}}, \tag{19}$$

where λ is given in Appendix B and $i = Ex, C, Co$.

Thus, the second optimization of W (see Appendix B) leads to

$$Max_2 W \approx W_{endo} - \frac{T_0}{\tau} N^2. \tag{20}$$

Furthermore, a third sequential optimization could be performed by considering the finite entropic action as a new constraint. This case is not developed here for brevity reasons.

4.2. Power Optimization in the Case of a Finite Heat Source (When G_H Is the Parameter)

The mean power of the modified Chambadal cycle for the condition of maximum work $Max_2 W$ is defined by

$$\overline{W}(Max_2 W) = \frac{W_{endo}}{\tau} - \frac{T_0}{\tau^2} N^2, \tag{21}$$

where $W_{endo} = G_H \left(\sqrt{T_{HS}} - \sqrt{T_0} \right)^2$ is the mechanical work output of the endo-reversible engine.

The power is maximized with respect to the cycle period τ. Thus, the expression of the optimum period is

$$\tau^* = \frac{2 T_0 N^2}{W_{endo}}. \tag{22}$$

This expression is analogous to the similar results obtained in [5], leading to

$$Max \overline{W} = \frac{W_{endo}{}^2}{4 T_0 N^2}. \tag{23}$$

The action of entropy production appearing in N diminishes the mean power of the engine. At the given endo-reversible work, the maximum power corresponds to the minimum of the N function, depending on the four entropy actions of the cycle, such that

$$N = \sqrt{\frac{T_0}{T_{HS}}C_{IH} + \sqrt{C_{IEx}} + \sqrt{C_{IC}} + \sqrt{C_{ICo}}}. \tag{24}$$

The main difference between Equation (23) and the previous results [5] comes from the imperfect heat transfer between the source and the converter in the Chambadal model.

5. Discussion

This paper proposed that the Special Issue *Carnot Cycle and Heat Engine Fundamentals and Applications II* completes the previous paper [12] published in Special Issue 1 and adds new results to a recently published paper [5].

Whatever variable is chosen for the modified Chambadal model work optimization (T_H or ΔS), the same optimum for the work per cycle is obtained with parameters G_H, T_{HS}, and T_0.

It appears that by introducing the duration of each transformation τ_i and the period of the cycle τ, the modified Chambadal model satisfies the Gouy-Stodola theorem. At the minimum of entropy production, the optimal durations are dependent on the transformation entropy actions. This result is new to our knowledge.

This new concept [5] allows a new subsequent sequential optimization. The optimal allocation of the entropy action coefficients is slightly different from the equipartition (a new form of the equipartition theorem [15,16]).

Thus, the fundamental aspect related to irreversibilities through the *new concept of entropy production action* seems promising. Furthermore, this new concept could contribute to the improvement of the global system analysis by conducting optimal dimension allocation. In this respect, finite physical dimensions analysis could be a complementary way to correlate with exergy analysis.

Further extensions of this work are foreseen in the near future.

6. Conclusions

Similarities and differences present in the literature regarding the optimization of energy, first law efficiency, and power of the modified Chambadal engine have been resituated and summarized since the publication of [12].

This approach allows for highlighting the evolution of the obtained results from the reversible Carnot engine case (thermostatics) to the endo-irreversible models related to the approaches of Novikov [6] and Ibrahim et al. [7] or to the entropy production method that we promote.

By generalizing a proposal from Esposito et al. [9] and defining the new concept of entropic action through a coefficient C_I (Js/K) for the entropy production of transformations all along the cycle, we achieved a new form of power optimization different from the one of Curzon and Ahlborn, since the internal converter irreversibilities and the heat transfer irreversibility between the heat source and converter were accounted for.

The maximum work per cycle was obtained for the irreversible cycle case. It depended on the entropic action coefficient of the four transformations of the cycle C_{Ii}, after which the power of the engine was sequentially optimized.

An optimal period of the cycle τ^* appeared, corresponding to the maximum mean power of the cycle. It generalized the recent published results [5] for a modified Chambadal engine.

This research continues to be developed by our team to obtain more general results.

Author Contributions: Conceptualization, M.F.; methodology, M.F.; software, M.C.; validation, M.F. and M.C.; formal analysis, M.F. and M.C.; writing—original draft preparation, M.F.; writing—review and editing, M.C.; visualization, M.F. and M.C.; supervision, M.F.; project administration, M.C. All authors have read and agreed to the published version of the manuscript.

Funding: This research received no external funding.

Institutional Review Board Statement: Not applicable.

Informed Consent Statement: Not applicable.

Conflicts of Interest: The authors declare no conflict of interest.

Appendix A. Work per Cycle of the Modified Chambadal Engine with the Entropy Production Method

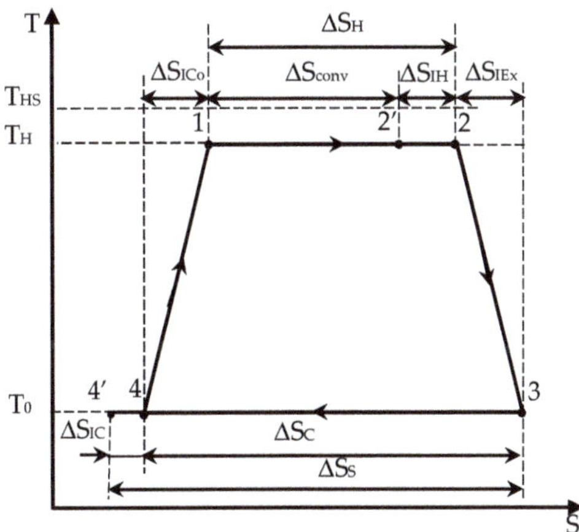

Figure A1. Carnot engine cycle with internal irreversibilities along the four transformations of the cycle, illustrated in a *T-S* diagram.

It results from Figure A1 that the various heats exchanged over the irreversible cycle (1–2–3–4) are expressed as follows:

- $Q_H = T_H \Delta S_H$ is the heat received by the cycled medium from the hot source (energy expense), corresponding to the heat transfer at the hot side;
- $Q_{conv} = T_H \Delta S_{conv}$, heat converted in mechanical energy during the isothermal process at T_H, with corresponding production of entropy ΔS_{IH} such that:

$$Q_{conv} = T_H(\Delta S_H - \Delta S_{IH}). \tag{A1}$$

- $Q_C = T_0 \Delta S_C$, where $\Delta S_C = \Delta S_S - \Delta S_{IC}$.

Note that ΔS_{IC} is the entropy production during the irreversible isotherm at T_0 and ΔS_S is the entropy rejected to the sink such that $Q_S = T_0 \Delta S_S$.

Thus, the entropy balance over the cycle is

$$\Delta S_{conv} + \Delta S_I = \Delta S_S \tag{A2}$$

The total entropy production over the cycle ΔS_I is represented by

$$\Delta S_I = \Delta S_{IH} + \Delta S_{IEx} + \Delta S_{IC} + \Delta S_{ICo}, \tag{A3}$$

where

ΔS_{IH} is the entropy production during the isothermal transformation at T_H, ΔS_{IE} is the entropy production during the adiabatic expansion from T_H to T_0, ΔS_{IC} is the entropy production during the isothermal transformation at T_0, and ΔS_{ICo}, is the entropy production during the adiabatic compression from T_0 to T_H.

The energy balance over the cycle for the system comprising the converter, the heat source, and the sink (with the source and sink as perfect thermostats) provides

$$W = Q_{conv} - Q_S. \tag{A4}$$

Various forms of mechanical energy are obtainable from this point by combining the preceding relations. Thus, one may express W as follows:

1. With ΔS_{conv} as the reference entropy:

$$W = T_H \Delta S_{conv} - T_0 \Delta S_S, \tag{A5}$$

$$W = (T_H - T_0)\Delta S_{conv} - T_0 \Delta S_I. \tag{A6}$$

2. With ΔS_S as the reference entropy:

$$W = (T_H - T_0)\Delta S_S - T_H \Delta S_I. \tag{A7}$$

3. With ΔS_S or ΔS_S as the reference entropy:

$$W = T_H(\Delta S_H - \Delta S_{IH}) - T_C(\Delta S_C + \Delta S_{IC}). \tag{A8}$$

We prefer to choose between Equations (A6) and (A7). Note that Equation (A7) was the one used by Esposito et al. [9].

We use Equation (A6) here because it gave back known results, particularly the Gouy-Stodola theorem, with ΔS_{conv} being a parameter. Thus, the maximum energy occurs when $\Delta S_I = 0$ such that

$$W_{endo} = (T_H - T_0)\Delta S_{conv}. \tag{A9}$$

This corresponds to the endo-reversible model of Chambadal.

In Section 3, we proposed a complete Chambadal model taking account entropy production all along the cycle.

Appendix B. Work Optimization Relative to Time (Frequency)

Using the Lagrange multipliers method with the frequencies $f_i = \frac{1}{\tau_i}$ as variables, we get the following function:

$$
\begin{aligned}
L(f_i) = {} & \left(\sqrt{G_H T_{HS}} - \sqrt{(G_H + C_{IH} f_H)T_0} \right)^2 \\
& - T_0(C_{IH} f_H + C_{IEx} f_{Ex} + C_{IC} f_C + C_{ICo} f_{Co}) \\
& - \lambda \left(\frac{1}{f_H} + \frac{1}{f_{Ex}} + \frac{1}{f_C} + \frac{1}{f_{Co}} - \tau \right).
\end{aligned}
\tag{A10}
$$

The vector of optimal values is

$$f_{Ex}^* = \sqrt{\frac{\lambda}{T_0 C_{IEx}}} \quad ; \quad f_C^* = \sqrt{\frac{\lambda}{T_0 C_{IC}}} \quad ; \quad f_{Co}^* = \sqrt{\frac{\lambda}{T_0 C_{ICo}}}, \tag{A11}$$

Additionally, the following is a non-linear equation to solve numerically for f_H^*:

$$f_H^2 = \lambda \sqrt{\frac{G_H + C_{IH} f_H}{G_H}} \frac{1}{\sqrt{T_{HS} T_0 C_{IH}}}. \tag{A12}$$

In the reasonable case of low irreversibility on the T_H isotherm ($C_{IH}f_H \ll G_H$), a good approximation of f_H^* is

$$f_H^* = \sqrt{\frac{\lambda}{\sqrt{T_{HS}T_0}C_{IH}}}.$$
(A13)

The finitude constraint on τ_i allows for determining the $\sqrt{\lambda}$ expression as

$$\sqrt{\lambda} = \frac{N\sqrt{T_0}}{\tau},$$
(A14)

where

$$N = \sqrt{\frac{T_0}{T_{HS}}}C_{IH} + \sqrt{C_{IEx}} + \sqrt{C_{IC}} + \sqrt{C_{ICo}}.$$
(A15)

Finally, we get

$$Max_2 W \approx W_{endo} - \frac{T_0}{\tau}N^2.$$
(A16)

References

1. Carnot, S. *Réflexion sur la Puissance Motrice du feu et des Machines Propres à Développer Cette Puissance*; Albert Blanchard: Paris, France, 1953. (In French)
2. Curzon, F.L.; Ahlborn, B. Efficiency of a Carnot engine at maximum power output. *Am. J. Phys.* **1975**, *43*, 22–24. [CrossRef]
3. Feidt, M. Carnot Cycle and Heat Engine Fundamentals and Applications. *Entropy* **2020**, *22*, 348. [CrossRef] [PubMed]
4. Feidt, M. The History and Perspectives of Efficiency at Maximum Power of the Carnot Engine. *Entropy* **2017**, *19*, 369. [CrossRef]
5. Feidt, M.; Feidt, R. Endo-irreversible thermo-mechanical Carnot engine with new concept of entropy production action coefficient. *Eur. Phys. J. Appl. Phys.* **2021**, *94*, 30901. [CrossRef]
6. Novikov, I. The efficiency of atomic power stations (a review). *J. Nucl. Energy* **1958**, *7*, 125–128. [CrossRef]
7. Ibrahim, O.M.; Klein, S.A.; Mitchell, J.W. Optimum Heat Power Cycles for Specified Boundary Conditions. *J. Eng. Gas Turbines Power* **1991**, *113*, 514–521. [CrossRef]
8. Gouy, G. Sur l'énergie utilizable. *J. Phys.* **1889**, *8*, 501–508. (In French)
9. Esposito, M.; Kawai, R.; Lindenberg, K.; Van den Broeck, C. Efficiency at Maximum Power of Low-Dissipation Carnot Engines. *Phys. Rev. Lett.* **2010**, *105*, 150603. [CrossRef] [PubMed]
10. Feidt, M.; Costea, M. From Finite Time to Finite Physical Dimensions Thermodynamics: The Carnot Engine and Onsager's Relations Revisited. *J. Non-Equilib. Thermodyn.* **2018**, *43*, 151–161. [CrossRef]
11. Dorfman, K.E.; Xu, D.; Cao, J. Efficiency at maximum power of a laser quantum heat engine enhanced by noise-induced coherence. *Phys. Rev. E* **2018**, *97*, 042120. [CrossRef] [PubMed]
12. Feidt, M.; Costea, M. Progress in Carnot and Chambadal Modeling of Thermomechanical Engine by Considering Entropy Production and Heat Transfer Entropy. *Entropy* **2019**, *21*, 1232. [CrossRef]
13. Barranco-Jimenez, M.A. Finite-time thermoeconomic optimization of a non endoreversible heat engine model. *Rev. Mex. Fís.* **2009**, *55*, 211–220.
14. Chambadal, P. *Les Centrales Nucléaires*; Armand Colin: Paris, France, 1957. (In French)
15. Tondeur, D. Optimisation Thermodynamique: Équipartition de Production d'entropie. Available online: https://hal.archives-ouvertes.fr/hal-00560251/ (accessed on 25 October 2021). (In French).
16. Tondeur, D. Optimisation Thermodynamique. Equipartition: Exemples et Applications. Available online: https://hal.archives-ouvertes.fr/hal-00560257/ (accessed on 25 October 2021). (In French).

Article

Performance Optimizations with Single-, Bi-, Tri-, and Quadru-Objective for Irreversible Diesel Cycle

Shuangshuang Shi [1,2], Lingen Chen [1,2,*], Yanlin Ge [1,2,*] and Huijun Feng [1,2]

[1] Institute of Thermal Science and Power Engineering, Wuhan Institute of Technology, Wuhan 430205, China; shishuangshuang20@163.com (S.S.); huijunfeng@139.com (H.F.)
[2] School of Mechanical & Electrical Engineering, Wuhan Institute of Technology, Wuhan 430205, China
* Correspondence: lgchenna@yahoo.com (L.C.); geyali9@hotmail.com (Y.G.)

Abstract: Applying finite time thermodynamics theory and the non-dominated sorting genetic algorithm-II (NSGA-II), thermodynamic analysis and multi-objective optimization of an irreversible Diesel cycle are performed. Through numerical calculations, the impact of the cycle temperature ratio on the power density of the cycle is analyzed. The characteristic relationships among the cycle power density versus the compression ratio and thermal efficiency are obtained with three different loss issues. The thermal efficiency, the maximum specific volume (the size of the total volume of the cylinder), and the maximum pressure ratio are compared under the maximum power output and the maximum power density criteria. Using NSGA-II, single-, bi-, tri-, and quadru-objective optimizations are performed for an irreversible Diesel cycle by introducing dimensionless power output, thermal efficiency, dimensionless ecological function, and dimensionless power density as objectives, respectively. The optimal design plan is obtained by using three solution methods, that is, the linear programming technique for multidimensional analysis of preference (LINMAP), the technique for order preferences by similarity to ideal solution (TOPSIS), and Shannon entropy, to compare the results under different objective function combinations. The comparison results indicate that the deviation index of multi-objective optimization is small. When taking the dimensionless power output, dimensionless ecological function, and dimensionless power density as the objective function to perform tri-objective optimization, the LINMAP solution is used to obtain the minimum deviation index. The deviation index at this time is 0.1333, and the design scheme is closer to the ideal scheme.

Keywords: irreversible Diesel cycle; power output; thermal efficiency; ecological function; power density; finite time thermodynamics

Citation: Shi, S.; Chen, L.; Ge, Y.; Feng, H. Performance Optimizations with Single-, Bi-, Tri-, and Quadru-Objective for Irreversible Diesel Cycle. *Entropy* **2021**, *23*, 826. https://doi.org/10.3390/e23070826

Academic Editor: Michel Feidt

Received: 25 May 2021
Accepted: 23 June 2021
Published: 28 June 2021

1. Introduction

As a further extension of traditional irreversible process thermodynamics, finite time thermodynamics [1–13] have been applied to analyze and optimize performances of actual thermodynamic cycles, and great progress has been made. The application of finite time thermodynamics to study the optimal performance of Diesel cycles represents a new technology for improving and optimizing Diesel heat engines, and a new method for studying Diesel cycles has been developed. Assuming the working fluid's specific heats are constants [14–24] and vary with its temperature [25–32], many scholars have studied the performance of irreversible Diesel cycles with various objective functions, such as power output (P), thermal efficiency (η), and ecological functions (E, which was defined as the difference between the exergy flow rate and the exergy loss).

In addition to the above objective functions, Sahin et al. [33,34] took power density (P_d, defined as the ratio of the cycle P to the maximum specific volume) as a new optimization criterion to optimize Joule–Brayton engines and found that the heat engine designed under the P_d criterion has higher η and a smaller size when no loss is considered. Chen et al. [35]

15

introduced the objective function P_d into the thermodynamic analysis and optimization of the Atkinson cycle. Atmaca and Gumus [36] compared and analyzed the optimal performance of a reversible Diesel cycle based on the P, P_d, and effective P (which was defined as the product of power output and thermal efficiency) criteria. Raman and Kumar [37] conducted thermodynamic analysis and optimization of a reversible Diesel cycle under the criteria of P, P_d, and effective P when the working fluid's specific heats were linearly functioning with temperature. Rai and Sahoo [38] analyzed the influences of different losses on the effective P, effective P_d, and total heat loss of an irreversible Diesel cycle when the working fluid's specific heats were non-linearly functioning with temperature. Gonca and Palaci [39] analyzed and compared design parameters under the effective P and effective P_d criteria of an irreversible Diesel cycle.

The research mentioned above only optimized a single-objective function and did not optimize multiple objective functions at the same time. Therefore, NSGA-II can be used to solve a multi-objective optimization (MOO) problem, and MOO can be performed for the combination of different objective functions.

Ahmadi et al. [40–43] carried out MOO for an irreversible radiant heat engine [40], fuel cell combined cycle [41,42], and Lenoir heat engine [43] with different objective functions. Shi et al. [44] and Ahmadi et al. [45] performed MOO of the Atkinson cycle when the working fluid's specific heats were constants [44] and varied with temperature non-linearly [45]. Gonzalez et al. [46] performed MOO on P, η, and entropy generation of an endoreversible Carnot engine and analyzed the stability of the Pareto frontier. Ata et al. [47] performed parameter optimization and sensitivity analysis for an organic Rankine cycle with a variable temperature heat source. Herrera et al. [48] and Li et al. [49] performed MOO of η and emissions of a regenerative organic Rankine cycle. Garmejani et al. [50] performed MOO of P, exergy efficiency, and investment cost for a thermoelectric power generation system. Tang et al. [51] and Nemogne et al. [52] performed MOO of an irreversible Brayton cycle [51] and an absorption heat pump cycle [52]. MOO has been applied for performance optimization of various processes and cycles [53–56].

Reference [24] established a relatively complete irreversible Diesel cycle model and studied the optimal performance of E. Firstly, based on the model established in the reference [24], this paper studies the optimal P_d performance of an irreversible Diesel cycle while considering the impacts of the cycle temperature ratio and three loss issues. Secondly, the maximum specific volume, maximum pressure ratio, and η are compared under the maximum P and maximum P_d criteria. Thirdly, applying NSGA-II with a compression ratio as the decision variable and cycle dimensionless P ($\overline{P}$, which is defined as P divided by maximum P), η, dimensionless P_d ($\overline{P}_d$, which is defined as P_d divided by maximum P_d), and dimensionless E ($\overline{E}$, which is defined as E divided by maximum E) as objective functions, the single-, bi-, tri-, and quadru-objective optimizations of an irreversible Diesel cycle are performed. Through three different solutions, that is, LINMAP, TOPSIS, and Shannon entropy, the deviation indexes obtained under different solutions are compared, and the optimized design scheme with the smallest deviation index is finally obtained.

2. Cycle Model

The working fluid is assumed to be an ideal gas. Figures 1 and 2 show the $T - s$ and $P - v$ diagrams of an irreversible Diesel cycle. It can be seen that $1 - 2$ is an adiabatic process, $2 - 3$ is a constant-pressure process, $3 - 4$ is an adiabatic process, and $4 - 1$ is a constant-volume process. The processes $1 - 2s$ and $3 - 4s$ are the isentropic and adiabatic processes, respectively.

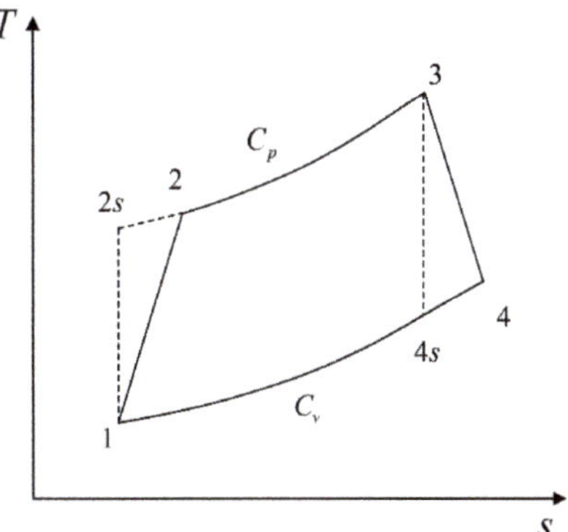

Figure 1. $T - s$ representation of the Diesel cycle.

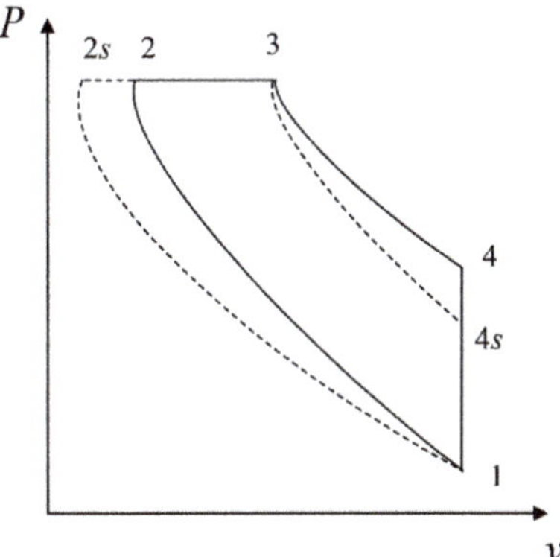

Figure 2. $P - v$ representation of the Diesel cycle.

The heat absorption and release rates are, respectively,

$$\dot{Q}_{in} = \dot{m}C_p(T_3 - T_2) \tag{1}$$

$$\dot{Q}_{out} = \dot{m}C_v(T_4 - T_1) \tag{2}$$

where $\dot{m}$ is the mass flow rate, and C_v and C_p are the specific heats under constant volume and pressure, respectively.

Some internal irreversibility loss (IIL) is caused by friction, turbulence, and viscous stress. The irreversible compression and expansion internal efficiencies are expressed as [16,19,20,30]

$$\eta_c = (T_{2s} - T_1)/(T_2 - T_1) \tag{3}$$

$$\eta_e = (T_3 - T_4)/(T_3 - T_{4s}) \tag{4}$$

The cycle compression ratio γ and temperature ratio τ are

$$\gamma = V_1/V_2 \tag{5}$$

$$\tau = T_3/T_1 \tag{6}$$

According to the property of isentropic process, one has

$$T_{2s} = T_1\gamma^{k-1} \tag{7}$$

$$(T_3/T_{2s})^k = T_{4s}/T_1 \tag{8}$$

According to Equations (3)–(8), one has

$$T_2 = T_1[(\gamma^{k-1} - 1)/\eta_c + 1] \tag{9}$$

$$T_{4s} = \tau^k T_1 / \gamma^{k(k-1)} \tag{10}$$

$$T_4 = T_1[\tau^k \eta_e / \gamma^{k(k-1)} - \tau\eta_e + \tau] \tag{11}$$

For the actual heat engine, there is heat transfer loss (HTL) between the working fluid and the cylinder. According to Refs. [14,24,27], it is known that the fuel exothermic rate is equal to the sum of the total endothermic rate and the HTL rate; one has

$$\dot{Q}_{leak} = A - \dot{Q}_{in} = B(T_3 + T_2 - 2T_0) \tag{12}$$

where A is the fuel exothermic rate and B is the HTL coefficient.

Similarly, as the piston generates friction with the cylinder wall when running at high speed, the friction loss (FL) of the cycle cannot be ignored. As a four-stroke heat engine, a Diesel heat engine has four strokes of intake, compression, expansion, and exhaust, and all of them produce FL. According to Refs. [24,32], for the treatment of FL in each stroke, the FL during compression and expansion is included in internal irreversible losses. According to Refs. [57–59], the piston motion resistance in the intake process is greater than that in the exhaust process. If the friction coefficient in the exhaust process is μ, the equivalent friction coefficient, which includes the pressure drop loss in the intake process, is 3μ. The friction coefficients on the exhaust and intake stroke are μ and 3μ, respectively. There is a linear relationship between friction force and speed: $f\mu = -\mu v = -\mu dx/dt$, where x is the piston displacement and μ is the FL coefficient. The power consumed due to FL during the exhaust and intake strokes can be derived as

$$P_\mu = dW_\mu/dt = 4\mu(dx/dt)^2 = 4\mu v^2 \tag{13}$$

For a Diesel cycle, the average speed of the piston in four reciprocating motions is

$$\bar{v} = 4Ln \tag{14}$$

where n is the rotating speed and L is the stroke length.

Therefore, the power consumed by cycle FL is

$$P_\mu = 4\mu(4Ln)^2 = 64\mu(Ln)^2 \tag{15}$$

The cycle P and η are, respectively,

$$P = \dot{Q}_{in} - \dot{Q}_{out} - P_\mu = \dot{m}[C_p(T_3 - T_2) - C_v(T_4 - T_1)] - 64\mu(Ln)^2 \tag{16}$$

$$\eta = \frac{P}{\dot{Q}_{in} + \dot{Q}_{leak}} = \frac{\dot{m}[C_p(T_3 - T_2) - C_v(T_4 - T_1)] - 64\mu(Ln)^2}{\dot{m}C_p(T_3 - T_2) + B(T_2 + T_3 - 2T_0)} \tag{17}$$

According to the definition of P_d in Refs. [33–35], the P_d is expressed as

$$P_d = P/v_4 \tag{18}$$

According to Refs. [38,39], the total volume v_t, stroke volume v_s, and gap volume v_c of the cycle are defined as

$$v_t = v_s + v_c \tag{19}$$

$$v_s = \pi d^2 L/4 \tag{20}$$

$$v_c = \pi d^2 L/4(\gamma - 1) \tag{21}$$

In the Diesel cycle, $v_t = v_{max} = v_1$, $v_c = v_2$. According to Equations (5) and (17)–(19), one has

$$P_d = P/v_{max} = P/v_t = 4(\gamma - 1)P/\pi d^2 L\gamma \tag{22}$$

According to Ref. [24], an irreversible Diesel cycle has four kinds of entropy generation due to FL, HTL, IIL, and exhaust stroke to the environment. The four entropy generation rates are expressed as

$$\sigma_q = B[1/T_0 - 2/(T_2 + T_3)](T_3 + T_2 - 2T_0) \tag{23}$$

$$\sigma_\mu = P_\mu/T_0 = 64\mu(Ln)^2/T_0 \tag{24}$$

$$\sigma_{2s \to 2} = \dot{m}\int_{T_{2s}}^{T_2} C_p dT/T = \dot{m}C_p \ln(T_2/T_{2s}) \tag{25}$$

$$\sigma_{4s \to 4} = \dot{m}\int_{T_{4s}}^{T_4} C_v dT/T = \dot{m}C_v \ln(T_4/T_{4s}) \tag{26}$$

$$\sigma_{pq} = \dot{m}\int_{T_1}^{T_4} C_v dT(1/T_0 - 1/T) = \dot{m}C_v[(T_4 - T_1)/T_0 + \ln(T_1/T_4)] \tag{27}$$

Therefore, the total entropy generation rate is

$$\sigma = \sigma_q + \sigma_\mu + \sigma_{2s \to 2} + \sigma_{4s \to 4} + \sigma_{pq} \tag{28}$$

According to the definition of E in Ref. [24], the E is expressed as

$$E = P - T_0\sigma \tag{29}$$

According to the processing method of Refs. [35,44], $\overline{P}$, $\overline{P}_d$, and $\overline{E}$ are respectively defined as

$$\overline{P} = P/P_{\max} \tag{30}$$

$$\overline{P}_d = P_d/(P_d)_{\max} \tag{31}$$

$$\overline{E} = E/E_{\max} \tag{32}$$

According to Equations (4), (9) and (11) and given the compression ratio γ, the initial cycle temperature T_1, and the cycle temperature ratio τ, by solving the temperatures at the 2, 3, and 4 state points, the corresponding numerical solutions of $\overline{P}$, η, $\overline{P}_d$, and $\overline{E}$ can be obtained.

3. Maximum Power Density Optimization

The working fluid is assumed to be an ideal gas. According to the nature of the air, $T_0 = 300\,\text{K}$, $T_1 = 350\,\text{K}$, $\dot{m} = 1\,\text{mol/s}$, $k = 1.4$, $C_v = 20.78\,\text{J}/(\text{mol} \cdot \text{K})$, and $\tau = 5.78 - 6.78$. According to Refs. [24,44], the cycle parameters are determined: $\gamma = 1 - 100$, $B = 2.2\,\text{W/K}$, $\mu = 1.2\,\text{kg/s}$, $L = 0.07\,\text{m}$ and $n = 30\,\text{s}^{-1}$.

The relationships between the objective functions ($\overline{P}_d$ and η) of an irreversible Diesel cycle and the cycle design parameters (the cycle temperature ratio, HTL, FL, and IIL) are shown in Figures 3–6. It can be noticed that the relationship between $\overline{P}_d$ and γ ($\overline{P}_d - \gamma$) is a parabolic-like one. When no loss is considered, the relationship between $\overline{P}_d$ and η ($\overline{P}_d - \eta$) is a parabolic-like one, and when there is loss, the relationship curve of $\overline{P}_d - \eta$ is a loop-shaped one.

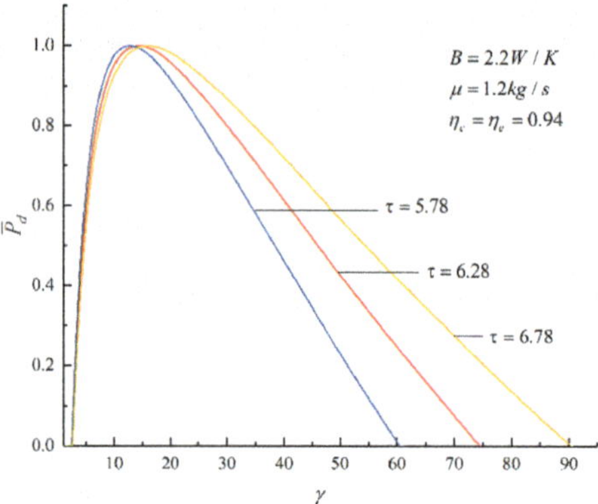

Figure 3. The effect of τ on $\overline{P}_d - \gamma$.

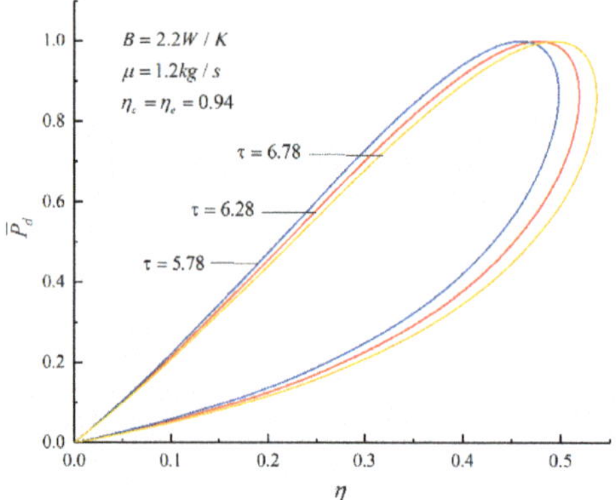

Figure 4. The effect of τ on $\overline{P}_d - \eta$.

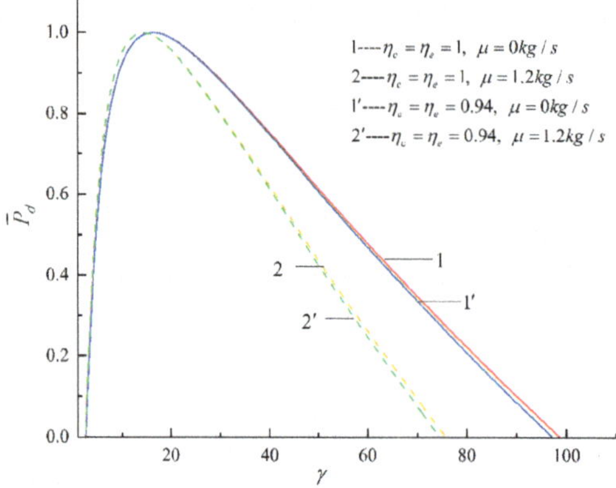

Figure 5. The effects of η_c, η_e, B, and b on $\overline{P}_d - \gamma$.

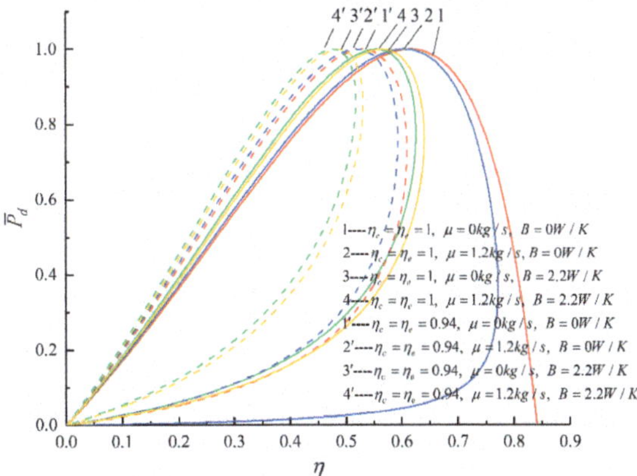

Figure 6. The effects of η_c, η_e, and b on $\overline{P}_d - \eta$.

Figures 3 and 4 show the effects of τ on the performances of $\overline{P}_d - \gamma$ and $\overline{P}_d - \eta$. According to Figure 3, it can be seen that there is an optimal compression ratio ($\gamma_{\overline{P}_d}$), which makes $\overline{P}_d$ reach the maximum. As τ increases, $\gamma_{\overline{P}_d}$ increases; when τ increases from 5.78 to 6.78, $\gamma_{\overline{P}_d}$ increases from 12.7 to 16 (an increase of 25.98%). According to Figure 4, there is thermal efficiency ($\eta_{\overline{P}_d}$) corresponding to the maximum $\overline{P}_d$. As τ increases, $\eta_{\overline{P}_d}$ increases; when τ increases from 5.78 to 6.78, $\eta_{\overline{P}_d}$ increases from 45.82% to 49.29% (an increase of 7.40%). It can be seen that with the increase in τ, $\gamma_{\overline{P}_d}$, and $\eta_{\overline{P}_d}$ corresponding to the maximum $\overline{P}_d$ also increases.

Figures 5 and 6 show the $\overline{P}_d - \gamma$ and $\overline{P}_d - \eta$ curves of the cycle when there are three different losses. Table 1 lists $\eta_{\overline{P}_d}$ when considering different losses and the percentage of the decrease in $\eta_{\overline{P}_d}$ compared with when no loss is considered. It can be seen that, with the increase in the losses considered, $\eta_{\overline{P}_d}$ decreases. When the three losses are considered at the same time, $\eta_{\overline{P}_d}$ decreases by 22.55% compared to that without any losses. According to Figure 5, it can be seen that as the compression ratio increases, $\overline{P}_d$ first increases and then decreases. According to Figure 6, it can be seen that when there are increases in HFL, FL, and IIL, $\eta_{\overline{P}_d}$ corresponding to the maximum $\overline{P}_d$ decreases.

Table 1. Comparison of the $\eta_{\overline{P}_d}$ in 8 cases.

Curve Number	Considered Loss	$\eta_{\overline{P}_d}$	Percentage of $\eta_{\overline{P}_d}$ Decrease
1	No loss	61.51%	0%
2	FL	60.36%	1.87%
3	HTL	56.45%	8.23%
4	FL and HTL	55.41%	9.92%
1′	IIL	52.97%	13.88%
2′	IIL and FL	51.84%	15.72%
3′	IIL and HTL	48.67%	20.87%
4′	IIL, HTL and FL	47.64%	22.55%

Figures 7–9 show the change trends of the corresponding maximum specific volume, maximum pressure ratio, and η with the τ under the maximum $\overline{P}$ and maximum $\overline{P}_d$ criteria of an irreversible Diesel cycle. According to Figures 7 and 8, compared with the corresponding results under the maximum $\overline{P}$ criterion, the maximum specific volume is smaller and the maximum pressure ratio is larger under the maximum $\overline{P}_d$ criterion. It is observed that the Diesel heat engine designed under the maximum $\overline{P}_d$ criterion has a smaller size.

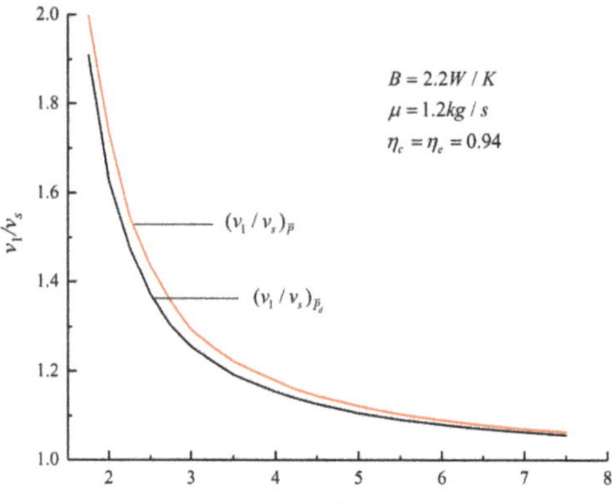

Figure 7. Variations of various v_1/v_s with τ.

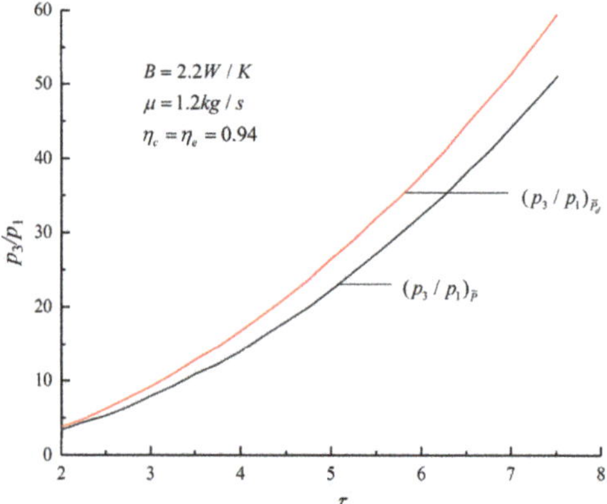

Figure 8. Variations of various p_3/p_1 with τ.

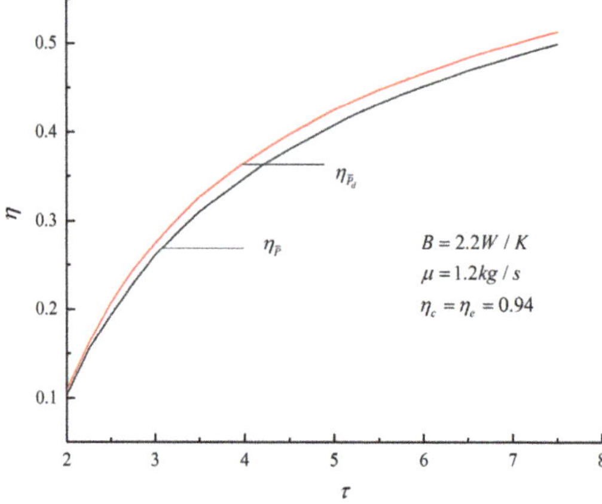

Figure 9. Variations of various η with τ.

According to Figure 9, the η of the cycle under the maximum $\overline{P}_d$ criterion is higher. When $\tau = 6.28$, the η obtained under the maximum $\overline{P}$ and maximum $\overline{P}_d$ criterion are 46.04%

and 47.64%, respectively. The latter is an increase of 3.54% over the former. Therefore, compared with the maximum $\overline{P}$ criterion, the engine designed under the maximum $\overline{P}_d$ criterion has a smaller size and a higher η.

4. Multi-Objective Optimization with Power Output, Thermal Efficiency, Ecological Function, and Power Density

MOO cannot make multiple objective functions reach the optimal value at the same time. The best compromise is achieved by comparing the pros and cons of each objective function. Therefore, the MOO solution set is not unique, and a series of feasible alternatives can be obtained, which are called Pareto frontiers. In this section, $\overline{P}$, η, $\overline{E}$, and $\overline{P}_d$ are used as objective functions; the compression ratio (γ) is used as an optimization variable; and NSGA-II [44–52] is used to perform bi-, tri-, and quadru-objective optimizations for an irreversible Diesel cycle. Through three different solutions, that is, LINMAP, TOPSIS, and Shannon entropy, the optimization results under different objective function combinations are obtained.

In the LINMAP solution, a minimum spatial distance from the ideal point is selected as the desired final optimal solution. In the TOPSIS solution, a maximum distance from the non-ideal point and a minimum distance from the ideal point are selected as the desired final optimal solution. In the Shannon entropy solution, a maximum value corresponding to a certain objective function is selected as the desired final optimal solution.

The optimization problems are solved with different optimization objective combinations, which form different MOO problems.

The six bi-objective optimization problems are as follows:

$$\max\left\{\begin{array}{c}\overline{P}(\gamma)\\\eta(\gamma)\end{array}\right. ,\max\left\{\begin{array}{c}\overline{P}(\gamma)\\\overline{E}(\gamma)\end{array}\right. ,\max\left\{\begin{array}{c}\overline{P}(\gamma)\\\overline{P}_d(\gamma)\end{array}\right. ,\max\left\{\begin{array}{c}\eta(\gamma)\\\overline{E}(\gamma)\end{array}\right. ,\max\left\{\begin{array}{c}\eta(\gamma)\\\overline{P}_d(\gamma)\end{array}\right. ,\max\left\{\begin{array}{c}\overline{E}(\gamma)\\\overline{P}_d(\gamma)\end{array}\right. \tag{33}$$

The four tri-objective optimization problems are as follows:

$$\max\left\{\begin{array}{c}\overline{P}(\gamma)\\\eta(\gamma)\\\overline{E}(\gamma)\end{array}\right. ,\max\left\{\begin{array}{c}\overline{P}(\gamma)\\\eta(\gamma)\\\overline{P}_d(\gamma)\end{array}\right. ,\max\left\{\begin{array}{c}\overline{P}(\gamma)\\\overline{E}(\gamma)\\\overline{P}_d(\gamma)\end{array}\right. ,\max\left\{\begin{array}{c}\eta(\gamma)\\\overline{E}(\gamma)\\\overline{P}_d(\gamma)\end{array}\right. \tag{34}$$

The one quadru-objective optimization problem is as follows:

$$\max\left\{\begin{array}{c}\overline{P}_d(\gamma)\\\eta(\gamma)\\\overline{E}(\gamma)\\\overline{P}_d(\gamma)\end{array}\right. \tag{35}$$

The evolution flow chart of NSGA-II is shown in Figure 10. The optimization results obtained by the combination of different objective functions in the three solutions are listed in Table 2. It can be seen that when single-objective optimization is performed under the criterions of maximum $\overline{P}, \eta$, $\overline{E}$, and $\overline{P}_d$, the deviation indexes (0.5828, 0.5210, 0.2086, and 0.4122, respectively) obtained are much larger than the result obtained by MOO. This indicates that the design scheme of MOO is more ideal. When taking $\overline{P}, \overline{E}$, and $\overline{P}_d$ as the optimization objectives to perform tri-objective optimization, the deviation index obtained by the LINMAP solution is smaller, and the design scheme is closer to the ideal scheme.

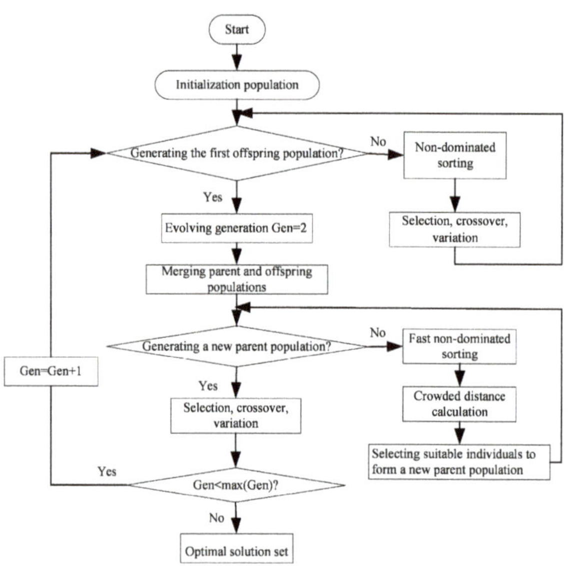

Figure 10. Flow chart of NSGA-II.

Figures 11–16 show the Pareto frontiers of bi-objective optimization ($\overline{P} - \eta$, $\overline{P} - \overline{E}$, $\overline{P} - \overline{P}_d$, $\eta - \overline{E}$, $\eta - \overline{P}_d$, and $\overline{E} - \overline{P}_d$). When $\overline{P}$ increases, η, $\overline{E}$, and $\overline{P}_d$ all decrease; when η increases, $\overline{E}$ and $\overline{P}_d$ both decrease; when $\overline{E}$ increases, $\overline{P}_d$ decreases. According to Table 1, when $\overline{P}$ and η or $\overline{P}$ and $\overline{E}$ are the objective functions, the deviation index obtained by the LINMAP solution is smaller. When $\overline{P}$ and $\overline{P}_d$ or η and $\overline{E}$ are the optimization objectives, the deviation index obtained by the Shannon entropy solution is smaller. When $\overline{E}$ and $\overline{P}_d$ are the optimization objectives, the deviation indexes obtained by the LINMAP and TOPSIS solutions are smaller than those obtained by the Shannon entropy solution. When η and $\overline{P}_d$ are the objective functions, the deviation index obtained by the TOPSIS solution is smaller.

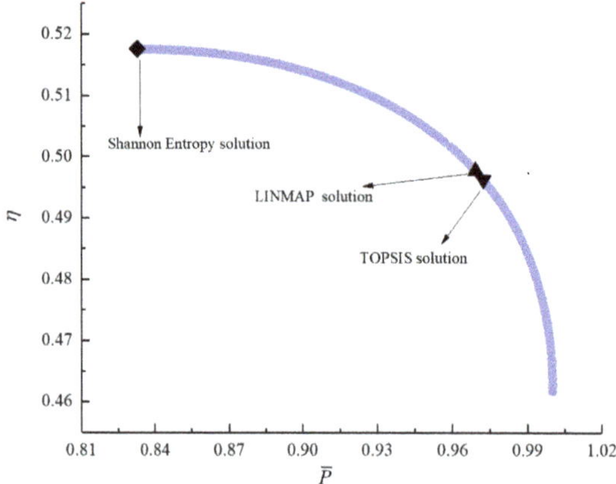

Figure 11. Bi-objective optimization on $\overline{P} - \eta$.

Table 2. Optimization results obtained by combining different objective functions.

Optimization Methods	Solutions	Optimization Variable	Optimization Objectives				Deviation Index
		γ	$\overline{P}$	η	$\overline{E}$	$\overline{P}_d$	D
Quadru-objective optimization $(\overline{P}, \eta, \overline{E}, \text{and } \overline{P}_d)$	LINMAP	18.0466	0.9615	0.5008	0.9809	0.9804	0.1342
	TOPSIS	18.0822	0.9611	0.5010	0.9815	0.9801	0.1346
	Shannon entropy	14.3437	0.9958	0.4769	0.8359	1.0000	0.4068
Tri-objective optimization $(\overline{P}, \eta, \text{and } \overline{E})$	LINMAP	18.2403	0.9591	0.5017	0.9842	0.9785	0.1366
	TOPSIS	18.5159	0.9556	0.5029	0.9882	0.9758	0.1422
	Shannon entropy	20.3584	0.9299	0.5095	1.0000	0.9545	0.2068
Tri-objective optimization $(\overline{P}, \eta, \text{and } \overline{P}_d)$	LINMAP	17.1965	0.9715	0.4966	0.9624	0.9878	0.1443
	TOPSIS	16.8933	0.9749	0.4949	0.9540	0.9900	0.1574
	Shannon entropy	14.3433	0.9958	0.4768	0.8359	1.0000	0.4068
Tri-objective optimization $(\overline{P}, \overline{E}, \text{and } \overline{P}_d)$	LINMAP	17.8459	0.9640	0.4999	0.9772	0.9823	0.1333
	TOPSIS	17.9598	0.9626	0.5004	0.9793	0.9812	0.1336
	Shannon entropy	14.3437	0.9958	0.4768	0.8359	1.0000	0.4068
Tri-objective optimization $(\eta, \overline{E}, \text{and } \overline{P}_d)$	LINMAP	18.7911	0.9520	0.5040	0.9916	0.9729	0.1495
	TOPSIS	18.7911	0.9520	0.5040	0.9916	0.9729	0.1495
	Shannon entropy	14.3437	0.9958	0.4769	0.8359	1.0000	0.4068
Bi-objective optimization $(\overline{P} \text{ and } \eta)$	LINMAP	17.4129	0.9691	0.4977	0.9678	0.9860	0.1380
	TOPSIS	17.3189	0.9722	0.4962	0.9655	0.9868	0.1384
	Shannon entropy	26.2726	0.8327	0.5176	0.9166	0.8647	0.5193
Bi-objective optimization $(\overline{P} \text{ and } \overline{E})$	LINMAP	18.0043	0.9620	0.5006	0.9802	0.9808	0.1339
	TOPSIS	18.2236	0.9593	0.5016	0.9839	0.9787	0.1364
	Shannon entropy	20.3584	0.9299	0.5095	1.0000	0.9545	0.2068
Bi-objective optimization $(\overline{P} \text{ and } \overline{P}_d)$	LINMAP	13.5850	0.9989	0.4699	0.7800	0.9989	0.5004
	TOPSIS	13.5850	0.9989	0.4699	0.7800	0.9989	0.5004
	Shannon entropy	14.3437	0.9958	0.4768	0.8359	1.0000	0.4068
Bi-objective optimization $(\eta \text{ and } \overline{E})$	LINMAP	21.6879	0.9097	0.5129	0.9948	0.9367	0.2645
	TOPSIS	21.6879	0.9097	0.5129	0.9948	0.9367	0.2645
	Shannon entropy	20.3584	0.9299	0.5095	1.0000	0.9545	0.2068
Bi-objective optimization $(\eta \text{ and } \overline{P}_d)$	LINMAP	18.4344	0.9566	0.5026	0.9871	0.9766	0.1403
	TOPSIS	18.1938	0.9597	0.5015	0.9834	0.9790	0.1359
	Shannon entropy	14.3437	0.9958	0.4768	0.8359	1.000	0.4068
Bi-objective optimization $(\overline{E} \text{ and } \overline{P}_d)$	LINMAP	18.5178	0.9555	0.5029	0.9882	0.9758	0.1422
	TOPSIS	18.5178	0.9555	0.5029	0.9882	0.9758	0.1422
	Shannon entropy	14.3437	0.9958	0.4769	0.8359	0.9999	0.4068
Maximum of $\overline{P}$	-	12.8106	1.0000	0.4617	0.7090	0.9952	0.5828
Maximum of η	-	26.2980	0.8323	0.5176	0.9160	0.8643	0.5210
Maximum of $\overline{E}$	-	20.4061	0.9293	0.5096	1.0000	0.9540	0.2086
Maximum of $\overline{P}_d$	-	14.3205	0.9960	0.4765	0.8330	1.0000	0.4122
Positive ideal point		-	1.0000	0.5176	1.0000	1.0000	-
Negative ideal point		-	0.8328	0.4618	0.7105	0.8647	-

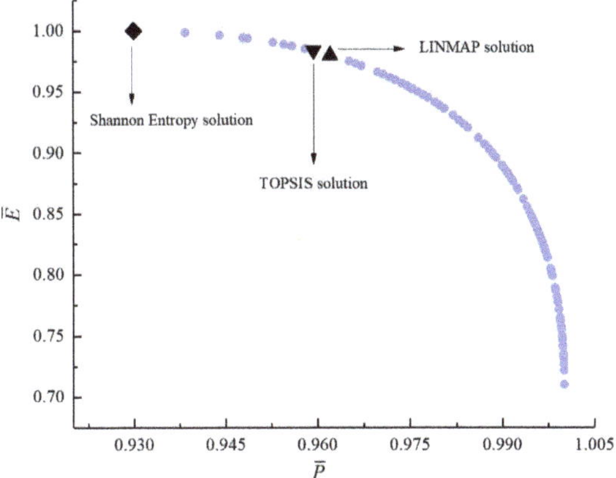

Figure 12. Bi-objective optimization on $\overline{P} - \overline{E}$.

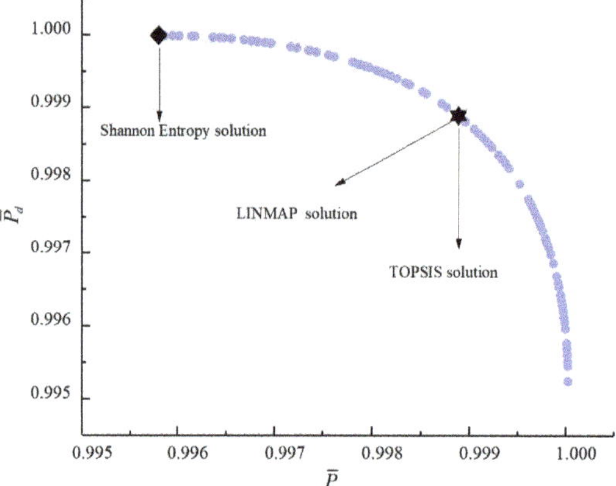

Figure 13. Bi-objective optimization on $\overline{P} - \overline{P}_d$.

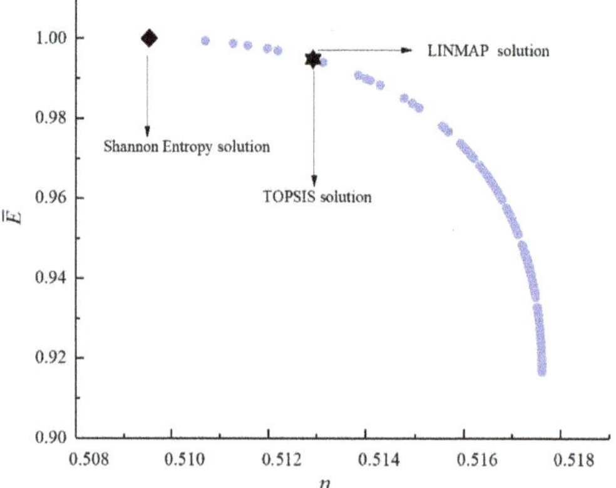

Figure 14. Bi-objective optimization on $\eta - \overline{E}$.

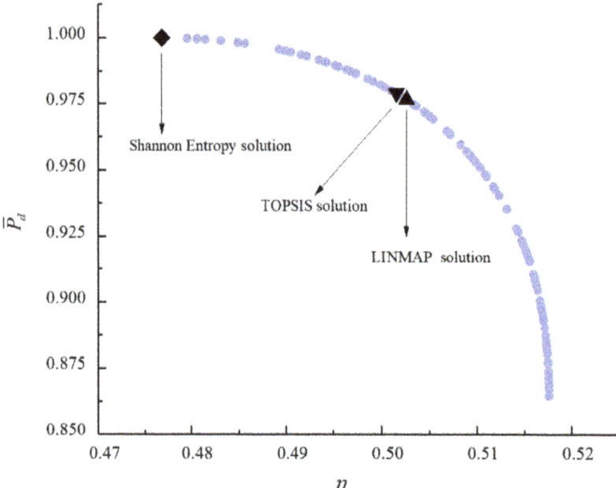

Figure 15. Bi-objective optimization on $\eta - \overline{P}_d$.

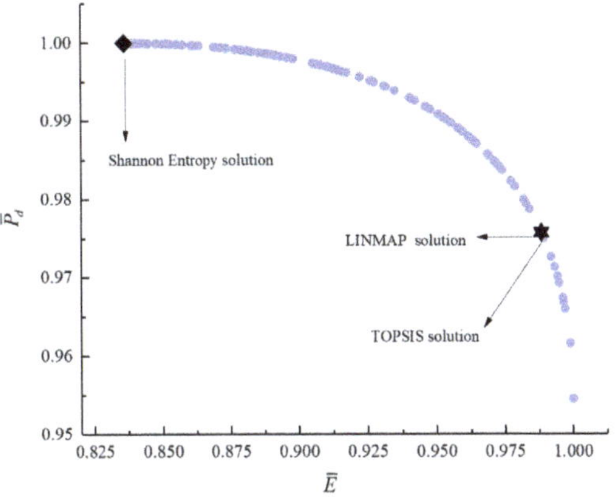

Figure 16. Bi-objective optimization on $\overline{E} - \overline{P}_d$.

Figures 17–20 show the Pareto frontiers of the tri-objective optimization ($\overline{P} - \eta - \overline{P}_d$, $\overline{P} - \eta - \overline{E}$, $\eta - \overline{E} - \overline{P}_d$, and $\overline{P} - \overline{E} - \overline{P}_d$). When $\overline{P}$ increases, η decreases, and $\overline{E}$ and $\overline{P}_d$ first increase and then decrease. When η increases, $\overline{P}_d$ decreases, and $\overline{E}$ first increases and then decreases. When η, $\overline{E}$, and $\overline{P}_d$ are the optimization objectives, the deviation indexes obtained by the LINMAP and TOPSIS solutions are smaller than those obtained by the Shannon entropy solution. When the combination of the other three objective functions are the optimization objectives, the deviation index obtained by the LINMAP solution is smaller, and the result is better.

Figure 21 shows the Pareto frontier of the quadru-objective optimization ($\overline{P} - \eta - \overline{E} - \overline{P}_d$). With the increase in $\overline{P}$, η increases, $\overline{P}_d$ decreases, and $\overline{E}$ first increases and then decreases. When $\overline{P}$, η, $\overline{E}$, and $\overline{P}_d$ are the optimization objectives, the deviation index obtained by the LINMAP solution is the smallest, and the result is the best.

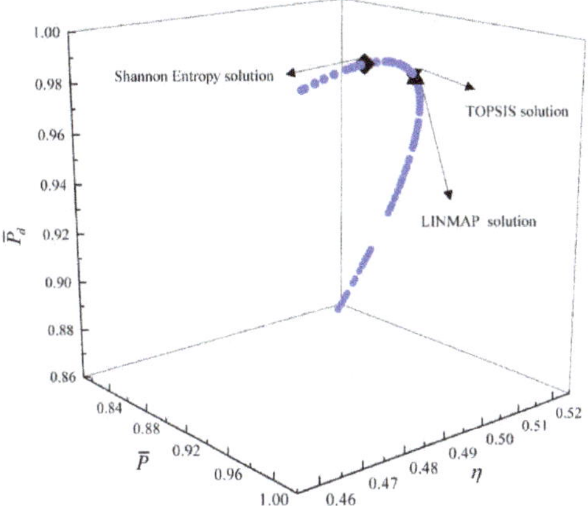

Figure 17. Tri-objective optimization on $\overline{P} - \eta - \overline{P}_d$.

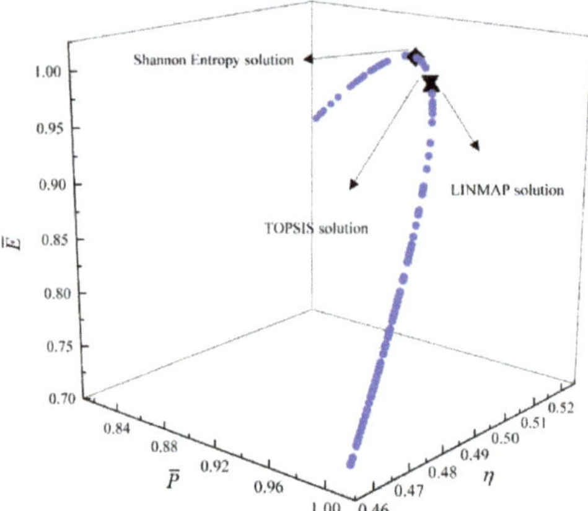

Figure 18. Tri-objective optimization on $\overline{P} - \eta - \overline{E}$.

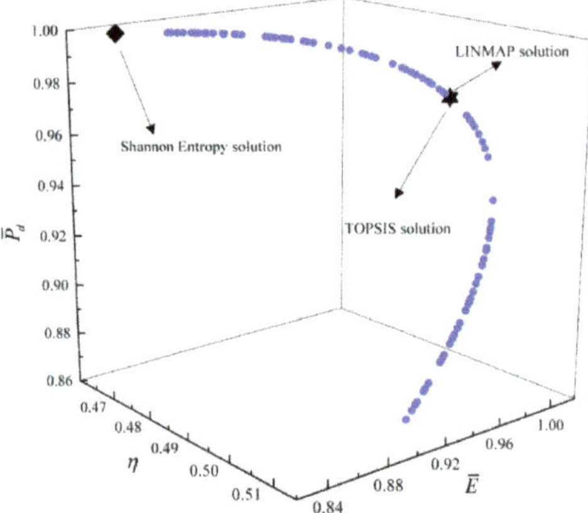

Figure 19. Tri-objective optimization on $\eta - \overline{E} - \overline{P}_d$.

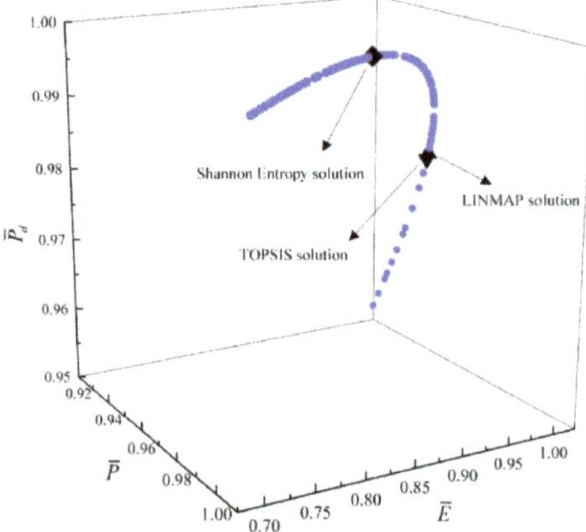

Figure 20. Tri-objective optimization on $\overline{P} - \overline{E} - \overline{P}_d$.

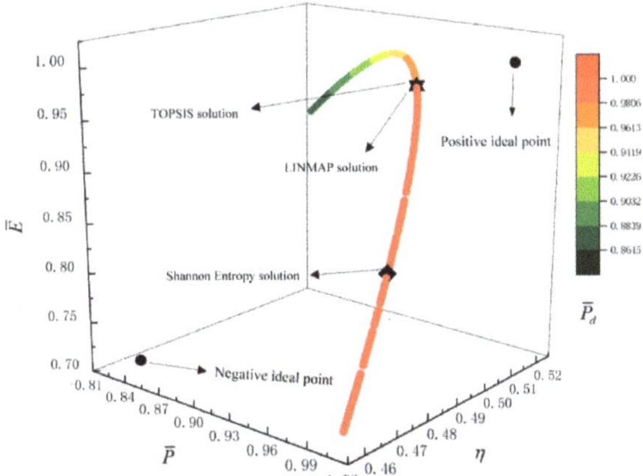

Figure 21. Quadru-objective optimization on $\overline{P} - \eta - \overline{E} - \overline{P}_d$.

5. Conclusions

The expression of the P_d of an irreversible Diesel cycle was derived in this paper, and the impacts of τ and three loss issues on the cycle of P_d versus γ and η characteristics were analyzed. The performance parameters (maximum specific volume, maximum pressure ratio, and η) of an irreversible Diesel cycle based on the criteria of maximum $\overline{P}$ and $\overline{P}_d$ were compared. Using three different solutions, including LINMAP, TOPSIS, and Shannon entropy, the results of single-, bi-, tri-, and quadru-objective optimization for an irreversible Diesel cycle were analyzed and compared. Comparing the deviation indexes obtained under different objective function combinations, the optimal design scheme was selected. The results showed the following:

1. The relationship curves of the cycles $\overline{P}_d - \gamma$ and $\overline{P}_d - \eta$ were a parabolic-like one and a loop-shaped one, respectively. With the increases in the cycle temperature ratio, the $\gamma_{\overline{P}_d}$ and $\eta_{\overline{P}_d}$ corresponding to the maximum $\overline{P}_d$ increased. With the increases in HFL, FL, and IIL, the $\gamma_{\overline{P}_d}$ and $\eta_{\overline{P}_d}$ corresponding to the maximum $\overline{P}_d$ decreased.
2. Under the maximum $\overline{P}_d$ criterion, a smaller size and higher efficiency engine will be designed.
3. The deviation index of MOO was smaller. When taking $\overline{P}$, $\overline{E}$, and $\overline{P}_d$ as the optimization objectives to perform tri-objective optimization, the deviation index obtained by the LINMAP solution was smaller, and the design scheme was closer to the ideal scheme.
4. The next step will be to use exergy efficiency optimization to further reinforce the results of MOO.

Author Contributions: Conceptualization, Y.G. and L.C.; funding acquisition, L.C.; methodology, S.S., L.C., Y.G. and H.F.; software, S.S., Y.G. and H.F.; supervision, L.C.; validation, S.S. and H.F.; writing—original draft, S.S. and Y.G.; writing—review and editing, L.C. All authors have read and agreed to the published version of the manuscript.

Funding: This paper is supported by The National Natural Science Foundation of China (Project No. 51779262) and Graduate Innovative Fund of Wuhan Institute of Technology (Project No. CX2020038).

Data Availability Statement: Data sharing not applicable.

Acknowledgments: The authors wish to thank the reviewers for their careful, unbiased, and constructive suggestions, which led to this revised manuscript.

Conflicts of Interest: The authors declare no conflict of interest.

Nomenclature

B	Heat transfer loss coefficient (W/K)
C_p	Specific heat at constant pressure (J/(mol · K))
C_v	Specific heat at constant volume (J/(mol · K))
$\overline{E}$	Dimensionless ecological function
$\overline{P}$	Dimensionless power output
$\overline{P}_d$	Dimensionless power density
Q	Heat transfer rate (W)
T	Temperature (K)
Greek symbols	
γ	Compression ratio (-)
η	Thermal efficiency (-)
μ	Friction coefficient (kg/s)
σ	Entropy generation rate (W/K)
τ	Temperature ratio (-)
Subscripts	
$\overline{P}_d$	Max power density condition
0	Environment
$1-4,2s,4s$	Cycle state points
Abbreviations	
FL	Friction loss
HTL	Heat transfer loss
IIL	Internal irreversibility loss
MOO	Multi-objective optimization

References

1. Andresen, B.; Berry, R.S.; Ondrechen, M.J.; Salamon, P. Thermodynamics for processes in finite time. *Acc. Chem. Res.* **1984**, *17*, 266–271. [CrossRef]
2. Chen, L.G.; Wu, C.; Sun, F.R. Finite time thermodynamic optimization or entropy generation minimization of energy systems. *J. Non Equilib. Thermodyn.* **1999**, *24*, 327–359. [CrossRef]
3. Andresen, B. Current trends in finite-time thermodynamics. *Angew. Chem. Int. Ed.* **2011**, *50*, 2690–2704. [CrossRef]
4. Feidt, M. Reconsideration of criteria and modeling in order to optimize the efficiency of irreversible thermomechanical heat engines. *Entropy* **2010**, *12*, 2470–2484. [CrossRef]
5. Dong, Y.; El-Bakkali, A.; Descombes, G.; Feidt, M.; Périlhon, C. Association of finite-time thermodynamics and a bond-graph approach for modeling an endoreversible heat engine. *Entropy* **2012**, *14*, 642–653. [CrossRef]
6. Ponmurugan, M. Attainability of maximum work and the reversible efficiency of minimally nonlinear irreversible heat engines. *J. Non Equilib. Thermodyn.* **2019**, *44*, 143–153. [CrossRef]
7. Dai, D.D.; Liu, Z.C.; Long, R.; Yuan, F.; Liu, W. An irreversible Stirling cycle with temperature difference both in non-isothermal and isochoric processes. *Energy* **2019**, *186*, 115875. [CrossRef]
8. Chen, L.G.; Feng, H.J.; Ge, Y.L. Power and efficiency optimization for open combined regenerative Brayton and inverse Brayton cycles with regeneration before the inverse cycle. *Entropy* **2020**, *22*, 677. [CrossRef] [PubMed]
9. Chen, J.F.; Li, Y.; Dong, H. Simulating finite-time isothermal processes with superconducting quantum circuits. *Entropy* **2021**, *23*, 353. [CrossRef] [PubMed]
10. Dobre, C.; Lavinia, G.L.; Alexandru, D.A.; Chisiu, G.; Constantin, M. Stirling refrigerating machine modeling using Schmidt and finite physical dimensions thermodynamic models: A comparison with experiments. *Entropy* **2021**, *23*, 368. [CrossRef] [PubMed]
11. Wang, R.B.; Ge, Y.L.; Chen, L.G.; Feng, H.J.; Wu, Z.X. Power and thermal efficiency optimization of an irreversible steady flow Lenoir cycle. *Entropy* **2021**, *23*, 425. [CrossRef]
12. Valencia-Ortega, G.; Levario-Medina, S.; Barranco-Jiménez, M.A. Local and global stability analysis of a Curzon-Ahlborn model applied to power plants working at maximum k-efficient power. *Phys. A Stat. Mech. Appl.* **2021**, *571*, 125863. [CrossRef]
13. Berry, R.S.; Salamon, P.; Andresen, B. How it all began. *Entropy* **2020**, *22*, 908. [CrossRef] [PubMed]
14. Chen, L.G.; Zen, F.M.; Sun, F.R.; Wu, C. Heat transfer effects on the net work output and power as function of efficiency for air standard Diesel cycle. *Energy* **1996**, *21*, 1201–1205. [CrossRef]
15. Parlak, A. The effect of heat transfer on performance of the Diesel cycle and exergy of the exhaust gas stream in a LHR Diesel engine at the optimum injection timing. *Energy Convers. Manag.* **2005**, *46*, 167–179. [CrossRef]
16. Parlak, A. Comparative performance analysis of irreversible Dual and Diesel cycles under maximum power conditions. *Energy Convers. Manag.* **2005**, *46*, 351–359. [CrossRef]
17. Zhao, Y.R.; Lin, B.H.; Zhang, Y.; Chen, J.C. Performance analysis and parametric optimum design of an irreversible Diesel heat engine. *Energy Convers. Manag.* **2006**, *47*, 3383–3392. [CrossRef]

18. Al-Hinti, I.; Akash, B.; Abu-Nada, E.; Al-Sarkhi, A. Performance analysis of air-standard Diesel cycle using an alternative irreversible heat transfer approach. *Energy Convers. Manag.* **2008**, *49*, 3301–3304. [CrossRef]
19. Ebrahimi, R. Performance optimization of a Diesel cycle with specific heat ratio. *J. Am. Sci.* **2009**, *5*, 59–63.
20. Zheng, S.Y.; Lin, G.X. Optimization of power and efficiency for an irreversible Diesel heat engine. *Front. Energy Power Eng. China* **2010**, *4*, 560–565. [CrossRef]
21. Rashidi, M.M.; Hajipour, A. Comparison of performance of air-standard Atkinson, Diesel and Otto cycles with constant specific heats. *Int. J. Adv. Des. Manuf. Technol.* **2013**, *6*, 57–62.
22. Zhu, F.L.; Chen, L.G.; Wang, W.H. Thermodynamic analysis and optimization of irreversible Maisotsenko-Diesel cycle. *Int. J. Therm. Sci.* **2019**, *28*, 659–668. [CrossRef]
23. Wu, H.; Ge, Y.L.; Chen, L.G.; Feng, H.J. Power, efficiency, ecological function and ecological coefficient of performance optimizations of an irreversible Diesel cycle based on finite piston speed. *Energy* **2021**, *216*, 119235. [CrossRef]
24. Ge, Y.L.; Chen, L.G.; Feng, H.J. Ecological optimization of an irreversible Diesel cycle. *Eur. Phys. J. Plus* **2021**, *136*, 1–13. [CrossRef]
25. Rocha-Martinez, J.A.; Navarrete-Gonzalez, T.D.; Pava-Miller, C.G. Otto and Diesel engine models with cyclic variability. *Rev. Mex. Fis.* **2002**, *48*, 228–234.
26. Al-Sarkhi, A.; Jaber, J.O.; Abu-Qudais, M.; Probert, S.D. Effects of friction and temperature-dependent specific-heat of the working fluid on the performance of a Diesel-engine. *Appl. Energy* **2006**, *83*, 153–165. [CrossRef]
27. Zhao, Y.R.; Chen, J.C. Optimum performance analysis of an irreversible Diesel heat engine affected by variable heat capacities of working fluid. *Energy Convers. Manag.* **2007**, *48*, 2595–2603. [CrossRef]
28. Ge, Y.L.; Chen, L.G.; Sun, F.R.; Wu, C. Performance of Diesel cycle with heat transfer, friction and variable specific heats of working fluid. *J. Energy Inst.* **2007**, *80*, 239–242. [CrossRef]
29. Ge, Y.L.; Chen, L.G.; Sun, F.R.; Wu, C. Performance of an endoreversible Diesel cycle with variable specific heats of working fluid. *Int. J. Ambient Energy* **2008**, *29*, 127–136. [CrossRef]
30. Ge, Y.L.; Chen, L.G.; Sun, F.R. Finite time thermodynamic modeling and analysis for an irreversible Diesel cycle. *Proc. Inst. Mech. Eng. Part D* **2008**, *222*, 887–894. [CrossRef]
31. Sakhrieh, A.; Abu-Nada, E.; Akash, B.; Al-Hinti, I.; Al-Ghandoor, A. Performance of diesel engine using gas mixture with variable specific heats model. *J. Energy Inst.* **2010**, *83*, 217–224. [CrossRef]
32. Hou, S.S.; Lin, J.C. Performance analysis of a Diesel cycle under the restriction of maximum cycle temperature with considerations of heat loss, friction, and variable specific heats. *Acta Phys. Pol. A* **2011**, *120*, 979–986. [CrossRef]
33. Sahin, B.; Kodal, A.; Yavuz, H. Efficiency of a Joule-Brayton engine at maximum power density. *J. Phys. D Appl. Phys.* **1995**, *28*, 1309. [CrossRef]
34. Sahin, B.; Kodal, A.; Yilmaz, T.; Yavuz, H. Maximum power density analysis of an irreversible Joule-Brayton engine. *J. Phys. D Appl. Phys.* **1996**, *29*, 1162. [CrossRef]
35. Chen, L.G.; Lin, J.X.; Sun, F.R.; Wu, C. Efficiency of an Atkinson engine at maximum power density. *Energy Convers. Manag.* **1998**, *39*, 337–341. [CrossRef]
36. Atmaca, M.; Gumus, M. Power and efficiency analysis of Diesel cycle under alternative criteria. *Arab. J. Sci. Eng.* **2014**, *39*, 2263–2270. [CrossRef]
37. Raman, R.; Kumar, N. Performance analysis of Diesel cycle under efficient power density condition with variable specific heat of working fluid. *J. Non-Equilib. Thermodyn.* **2019**, *44*, 405. [CrossRef]
38. Rai, R.K.; Sahoo, R.R. Effective power and effective power density analysis for water in diesel emulsion as fuel in Diesel engine performance. *Energy* **2019**, *180*, 893–902. [CrossRef]
39. Gonca, G.; Palaci, Y. Performance investigation into a Diesel engine under effective effciency-power-power density conditions. *Sci. Iran. B* **2019**, *26*, 843–855.
40. Ahmadi, M.H.; Ahmadi, M.A. Thermodynamic analysis and optimisation of an irreversible radiative-type heat engine by using non-dominated sorting genetic algorithm. *Int. J. Ambient Energy* **2016**, *37*, 403–408. [CrossRef]
41. Ahmadi, M.H.; Jokar, M.A.; Ming, T.Z.; Feidt, M.; Pourfayaz, F.; Astaraei, F.R. Multi-objective performance optimization of irreversible molten carbonate fuel cell-Braysson heat engine and thermodynamic analysis with ecological objective approach. *Energy* **2018**, *144*, 707–722. [CrossRef]
42. Ahmadi, M.H.; Sameti, M.; Sourkiaei, S.M.; Ming, T.Z.; Pourfayaz, F.; Chamkha, A.J.; Oztop, H.F.; Jokar, M.A. Multi-objective performance optimization of irreversible molten carbonate fuel cell-Stirling heat engine-reverse osmosis and thermodynamic assessment with ecological objective approach. *Energy Sci. Eng.* **2018**, *6*, 783–796. [CrossRef]
43. Ahmadi, M.H.; Nazari, M.A.; Feid, M. Thermodynamic analysis and multi-objective optimisation of endoreversible Lenoir heat engine cycle based on the thermo-economic performance criterion. *Int. J. Ambient Energy* **2019**, *40*, 600–609. [CrossRef]
44. Shi, S.S.; Ge, Y.L.; Chen, L.G.; Feng, H.J. Four-objective optimization of irreversible Atkinson cycle based on NSGA-II. *Entropy* **2020**, *22*, 1150. [CrossRef]
45. Ahmadi, M.H.; Pourkiaei, S.M.; Ghazvini, M.; Pourfayaz, F. Thermodynamic assessment and optimization of performance of irreversible Atkinson cycle. *Iran. J. Chem. Chem. Eng.* **2020**, *39*, 267–280.
46. Gonzalez-Ayala, J.; Roco, J.M.M.; Medina, A.; Calvo, H.A. Optimization, stability, and entropy in endoreversible heat engines. *Entropy* **2020**, *22*, 1323. [CrossRef]

47. Ata, S.; Kahraman, A.; Şahin, R. Prediction and sensitivity analysis under different performance indices of R1234ze ORC with Taguchi's multi-objective optimization. *Case Stud. Therm. Eng.* **2020**, *22*, 100785. [CrossRef]
48. Herrera-Orozco, I.; Valencia-Ochoa, G.; Duarte-Forero, J. Exergo-environmental assessment and multi-objective optimization of waste heat recovery systems based on Organic Rankine cycle configurations. *J. Clean. Prod.* **2021**, *288*, 125679. [CrossRef]
49. Li, Y.Y.; Li, W.Y.; Gao, X.Y.; Ling, X. Thermodynamic analysis and optimization of organic Rankine cycles based on radial-inflow turbine design. *Appl. Therm. Eng.* **2021**, *184*, 116277. [CrossRef]
50. Garmejani, H.A.; Hossainpou, S.H. Single and multi-objective optimization of a TEG system for optimum power, cost and second law efficiency using genetic algorithm. *Energy Convers. Manag.* **2021**, *228*, 113658. [CrossRef]
51. Tang, C.Q.; Chen, L.G.; Feng, H.J.; Ge, Y.L. Four-objective optimization for an irreversible closed modified simple Brayton cycle. *Entropy* **2021**, *23*, 282. [CrossRef] [PubMed]
52. Nemogne, R.L.F.; Wouagfack, P.A.N.; Nouadje, B.A.M.; Tchinda, R. Multi-objective optimization and analysis of performance of a four-temperature-level multi-irreversible absorption heat pump. *Energy Convers. Manag.* **2021**, *234*, 113967. [CrossRef]
53. Chen, L.G.; Tang, C.Q.; Feng, H.J.; Ge, Y.L. Power, efficiency, power density and ecological function optimizations for an irreversible modified closed variable-temperature reservoir regenerative Brayton cycle with one isothermal heating process. *Energies* **2020**, *13*, 5133. [CrossRef]
54. Zhang, L.; Chen, L.G.; Xia, S.J.; Ge, Y.L.; Wang, C.; Feng, H.J. Multi-objective optimization for helium-heated reverse water gas shift reactor by using NSGA-II. *Int. J. Heat Mass Transf.* **2020**, *148*, 119025. [CrossRef]
55. Sun, M.; Xia, S.J.; Chen, L.G.; Wang, C.; Tang, C.Q. Minimum entropy generation rate and maximum yield optimization of sulfuric acid decomposition process using NSGA-II. *Entropy* **2020**, *22*, 1065. [CrossRef]
56. Wu, Z.X.; Feng, H.J.; Chen, L.G.; Ge, Y.L. Performance optimization of a condenser in ocean thermal energy conversion (OTEC) system based on constructal theory and multi-objective genetic algorithm. *Entropy* **2020**, *22*, 641. [CrossRef]
57. Mozurkewich, M.; Berry, R.S. Finite-time thermodynamics: Engine performance improved by optimized piston motion. *Proc. Natl. Acad. Sci. USA* **1981**, *78*, 1986–1988. [CrossRef]
58. Mozurkewich, M.; Berry, R.S. Optimal paths for thermodynamic systems. The ideal Otto cycle. *J. Appl. Phys.* **1982**, *53*, 34–42. [CrossRef]
59. Hoffmann, K.H.; Watowich, S.J.; Berry, R.S. Optimal paths for thermodynamic systems. The ideal Diesel cycle. *J. Appl. Phys.* **1985**, *58*, 2125–2134. [CrossRef]

Article

Optimization Modeling of Irreversible Carnot Engine from the Perspective of Combining Finite Speed and Finite Time Analysis

Monica Costea [1,*], Stoian Petrescu [1], Michel Feidt [2], Catalina Dobre [1] and Bogdan Borcila [1]

[1] Department of Engineering Thermodynamics, University POLITEHNICA of Bucharest, Splaiul Independentei 313, 060042 Bucharest, Romania; stoian.petrescu@yahoo.com (S.P.); catalina.dobre@upb.ro (C.D.); bbd1188@yahoo.com (B.B.)

[2] Laboratory of Energetics, Theoretical and Applied Mechanics (LEMTA), URA CNRS 7563, University of Lorraine, 54518 Vandoeuvre-lès-Nancy, France; michel.feidt@univ-lorraine.fr

* Correspondence: monica.costea@upb.ro; Tel.: +40-021-402-9339

Abstract: An irreversible Carnot cycle engine operating as a closed system is modeled using the Direct Method and the First Law of Thermodynamics for processes with Finite Speed. Several models considering the effect on the engine performance of external and internal irreversibilities expressed as a function of the piston speed are presented. External irreversibilities are due to heat transfer at temperature gradient between the cycle and heat reservoirs, while internal ones are represented by pressure losses due to the finite speed of the piston and friction. Moreover, a method for optimizing the temperature of the cycle fluid with respect to the temperature of source and sink and the piston speed is provided. The optimization results predict distinct maximums for the thermal efficiency and power output, as well as different behavior of the entropy generation per cycle and per time. The results obtained in this optimization, which is based on piston speed, and the Curzon–Ahlborn optimization, which is based on time duration, are compared and are found to differ significantly. Correction have been proposed in order to include internal irreversibility in the externally irreversible Carnot cycle from Curzon–Ahlborn optimization, which would be equivalent to a unification attempt of the two optimization analyses.

Keywords: irreversible Carnot engine; optimization; thermodynamics with finite speed; internal and external irreversibilities; entropy generation calculation; thermodynamics in finite time

Citation: Costea, M.; Petrescu, S.; Feidt, M.; Dobre, C.; Borcila, B. Optimization Modeling of Irreversible Carnot Engine from the Perspective of Combining Finite Speed and Finite Time Analysis. *Entropy* **2021**, 23, 504. https://doi.org/10.3390/e23050504

Academic Editor: Peter Salamon

Received: 19 March 2021
Accepted: 18 April 2021
Published: 22 April 2021

Publisher's Note: MDPI stays neutral with regard to jurisdictional claims in published maps and institutional affiliations.

1. Introduction

Recent work [1] has emphasized that an analysis using the *finite time of the process* rather convey to a "physical potential optimization" than to an "engineering optimization" of thermal machine [2]. What is called *physical optimization* could provide more realistic performance compared to reversible Carnot cycle one, but it is still overvalued with respect to the actual one. Thus, the results of the physical optimization can be considered as upper bounds for real machine performance [3–5].

Moreover, criticisms have been addressed [6–11] to the results of Finite Time Thermodynamics (FTT) analysis of thermal machines, claiming that it failed to keep the promises, at least from the engineer's point of view. The main reason is the fact that FTT does not consider the internal losses generated by irreversibilities on a *fundamental basis*, since they have been introduced through a constant coefficient [12], factor of non-endoreversibility [13], degree of internal irreversibility [14], entropy variation ratio [15], ratio of two entropy differences [16], or entropy generation term as a function of temperature [17,18]. Therefore, the studies based on FTT approach cannot be effectively used by engineers for a better design and optimization study, leading to the conception and build of more efficient thermal machines since to apply optimization in a thermodynamic analysis, it needs to advance to the higher phases of the system design than the one based on endoreversibility assumption that is considered very early [10]. Furthermore, the internal irreversibilities contributed by the system components are inherently interconnected with external irreversibilities in real operation conditions, so the performance reported by FTT analysis may be even smaller compared to that of a real system [8].

These criticisms did not remain without reply [19–23]. Thus, some authors of the anti-criticism papers addressed the clarification of finite-time thermodynamics objectives and their inclusion in the efforts to approach the irreversible systems and their performance [21]. Others emphasized the meaning of time for thermodynamic processes, namely that of providing bounds by discussing nine general principles for finding bounds on the effectiveness of energy conversion [22] or bounds relative to the efficiency versus maximum power efficiency of heat engines [23].

However, regarding the usefulness of the FTT, the endoreversible model has the merit of launching nowadays the competition of finding new upper bounds of thermal machines performance, closer to the real one. Thus, progress has been made in the modeling and optimization of thermodynamic processes and cycles [24–32], with special attention to the common ones in thermal machines: Otto cycle [27], Stirling engine [28], Kalina cycle [30], and Brayton cycle [31,32]. The results obtained [30,31] have shown that besides the gains of FTT optimization with three or four objectives, the original results reported in the initial work of the FTT theory [3–5] are also revealed.

The engineering optimization is mainly concerned about internal irreversibility assessment by insight in dissipation mechanism, to approach and model the irreversible cycle performance. Both internal and external irreversibility are considered, conveying an actual optimization of thermal machine performance.

Although there is no operational Carnot machine, much has been written on the optimization of Carnot cycle, and in particular, on the heat engine cycle, endoreversible [33–39] or with internal and external irreversibilities [40–61]. One reason could be that the performance of the Carnot cycle represents upper bounds for actual operating machines. However, only in the 1990s was attention focused on analysis of the Carnot cycle that also includes internal irreversibilities [12,16–18,41,42,46–49].

The Thermodynamics with Finite Speed (TFS) has been shown to be able to provide analytical evaluation of internal irreversibilities in several machines (Stirling, Otto, Diesel, Brayton, Carnot) [60–68] and electrochemical devices [69], as a function of the speed of the piston. Actually, the finite speed of the piston (and process implicitly) is also responsible of external irreversibilities, namely the finite heat transfer rate from source to cycle fluid and then to sink. The computation scheme developed in TFS using the Direct Method is based on the *First Law of Thermodynamics for Processes with Finite Speed* that contains the main internal irreversibility causes of thermal machines expressed as a function of the average piston speed. By integration of the new expression of the First Law on each cycle process, analytical expression for performance (Power and Efficiency) is provided. It can be used to optimize theoretical cycles of actual thermal machines and most importantly, it was validated for 12 performing Stirling Engines (in 16 operational regimes) [63,64] and 4 Solar Stirling Motors [49,50].

In recent publications [54–58], it has been mentioned that only *Thermodynamics with Finite Speed* (TFS) developed the necessary tools to optimize thermal machines by considering internal losses in addition to external ones by analytical means. Based on these statements, it was concluded that using the above-mentioned achievements of TFS in combination with FTT tools could convey a more realistic and efficient approach of thermal machines.

The analytical approach relative to this combination is presented here by original models introducing irreversibilities step by step and leading to important results that are more accurate than those obtained by each irreversible thermodynamics branch separately.

Firstly, a brief presentation of the Curzon–Ahlborn modeling of an endoreversible Carnot engine is given, together with the discussion relative to the presence of the nice radical in other works.

Then, optimization models for a Carnot cycle engine in a closed system that operates with finite speed of the piston are presented. The speed is considered constant and equal to the average speed of the piston that moves with a classical rod–crankshaft mechanism; by using the First Law of Thermodynamics for Processes with Finite Speed and the Direct Method, the optimization analysis of this cycle with external and internal irreversibilities is developed. Heat losses between the two heat reservoirs temperature level through the engine are considered. External irreversibilities are due to the finite heat transfer rate at the source and sink are modeled by an irreversible coefficient added to the classical expression of heat transfer on isothermal process. Internal irreversibilities are included in the mathematical expression of the First Law of Thermodynamics for Processes with Finite Speed as non-dimensional pressure losses due to the non-uniformity of the fluid pressure in the cylinder and friction. The piston speed for maximum power and for maximum efficiency is found for a particular set of engine parameters and it is shown that the minimum entropy generation per cycle occurs at maximum power. This analysis provides lower values of Carnot cycle efficiency than predicted by the Curzon–Ahlborn approach that was considered for comparison.

A further development of the model aims to combine the analysis of the Carnot cycle engine with only external irreversibility from Finite Time Thermodynamics (FTT) with the main advantage of the Thermodynamics with Finite Speed (TFS) approach, namely the internal irreversibility quantification as a function of the speed of the process (piston). Thus, corrections of the power output, efficiency, and optimized cycle fluid temperature in FTT optimization results based on the calculated speed of processes from the duration time in FTT and average piston speed in TFS. It results that when internal ireversibilities (speeds and friction) are included, the performance predicted by a TFS analysis is better than that predicted by an FTT analysis.

The first unification attempt between TFS and FTT considers only pressure losses due to the non-uniformity of the pressure in the cylinder as a function of piston speed. The analytical development of the model provides modified Curzon–Ahlborn expression for the externally irreversible Carnot cycle to also include the internal irreversibility. Equations for the optimum cycle temperature, maximum power, and efficiency for the internally and externally irreversible cycle are presented. The corrections are shown to increase with increased piston speed and to be significant at high but realizable piston speeds. The optimum temperature corresponding to maximum power is shown to increase with increased piston speed.

Then, a further step in the unification attempt between TFS and FTT is done by considering in addition to the Finite Speed, two other causes of internal irreversibility given by friction and throttling. Thus, based on the first unification achievement, new expressions are derived for the power output and efficiency of the direct Carnot cycle with finite speed processes. The results emphasize optimum speed values generating maximum power output, as well as the effect of irreversibilities on the optimum high temperature of the cycle.

The overview on the results of these models emphasizes that a significant difference exists between the results of the two optimization analyses in the sense that FTT optimization seems to be an upper bound when compared to the engineering optimization based on TFS and the Direct Method.

2. Optimization Models of Carnot Cycle Engine

2.1. Models in Thermodynamics in Finite Time Analysis Seeking for Maximum Power Output of Carnot Cycle Engine

The Curzon–Ahlborn modeling of the Carnot-type engine [3] refers to a cycle that is internally reversible but with no thermal equilibrium between the working fluid and the thermal reservoirs during the isothermal heat input and heat rejection, respectively. Furthermore, there exists a finite time duration of heat transfer given by Newton's heat transfer law during the isothermal processes. The expression of the power output of the

Curzon and Ahlborn cycle allows a maximum for which the corresponding efficiency is given by what was called nice radical.

Actually, the efficiency of a Carnot engine is treated for the case where the power output is limited by the rates of heat transfer to and from the working substance. It is shown that the efficiency, η_{CA}, at maximum power output is given by the expression $\eta_{CA} = 1 - (T_2/T_1)^{1/2}$ where T_1 and T_2 are the respective temperatures of the heat source and heat sink. It results in an efficiency less than the one introduced by Carnot ($\eta = 1 - (T_2/T_1)$), and it is shown that the existing engines performance is well described by the above result.

Before the Curzon and Ahlborn analysis, a similar approach aiming to maximize the power output and the nice radical has appeared in Chambadal modeling of the Carnot engine [4], but its model used heat capacity rate instead of heat conductances.

Almost at the same time, Novikov [5] has also found the nice radical.

The above-mentioned models and mainly the Curzon–Ahlborn one, which remain as references for the Carnot machine optimization in the frame of what was called Thermodynamics in Finite Time.

2.2. Models of Irreversible Carnot Cycle Engine in Thermodynamics with Finite Speed

2.2.1. First Law of Thermodynamics for Processes with Finite Speed in Closed System

The optimization modeling presented in this section proceeds from a basis of thermodynamic fundamentals, systematically detailed and developed, starting from a unique equation called the *First Law of Thermodynamics for Processes with Finite Speed* [59,70–79]. The advantages of using this equation instead of the one from Classical Reversible Thermodynamics consists of its capability to account for both causes and mechanisms of irreversibility generation in complex cycles or real machines such as Stirling Engines, as well as in other cycles such as Otto, Diesel, Brayton, and Carnot cycles [60,71–73]. In addition, it is capable to consider both internal and external irreversibilities.

By integrating this equation for irreversible process step by step on each transformation of the cycle, the efficiency and power output are determined *analytically*. These expressions contain the *causes of irreversibility*, namely, the *finite speed of the piston*, an important parameter that can be optimized, for *Maximum Efficiency* or *Maximum Power*.

The mathematical expression of the First Law of Thermodynamics for Processes with Finite Speed in a closed system in its differential form is [59,70–76,78]:

$$dU = \delta Q - p_{av,i}\left(1 \pm \frac{aw}{c} \pm \frac{f \cdot \Delta p_f}{p_{av,i}}\right)dV, \tag{1}$$

and the irreversible work for these processes [59,70–76,78]:

$$\delta W_{irrev} = p_{av,i}\left(1 \pm \frac{aw}{c} \pm \frac{\Delta p_f}{p_{av,i}}\right)dV \tag{2}$$

where U—internal energy, Q—heat, W—mechanical work, $p_{av,i}$—instantaneous average pressure of the gas, w—average speed of the piston, c—average molecular speed, Δp_f—pressure losses due to friction, a—coefficient depending the gas nature, f—coefficient relative to the amount of heat generated by friction that remains in the cycle, and V—volume.

In the previous equations, the plus sign corresponds to the compression processes and the minus sign corresponds to the expansion ones.

Regarding the terms appearing in the right member, the first term in the parenthesis accounts for the irreversibility generated by the Finite Speed of the piston, w, and due to the non-uniformity of the pressure in the cylinder. Therefore, the pressure on the piston p_p is larger during compression and smaller during expansion than the pressure on the head of the cylinder p_c, and this is also the case for the instantaneous average pressure in the gas $p_{av,i}$ [47,59–61,76]. The experimental verification of this term is described in references [51,59–61]. The second term in the parenthesis takes into account the irreversibility generated by the friction between moving parts of the machine (piston–cylinder, bearings, etc.) [47,60,61]. When the processes in the machine involve internal throttling, a third term is added in the First Law for Processes with Finite Speed [47,60,61], playing an important role in the optimization of Stirling machines [51,59–67,77,80]. This term is less important in the Carnot cycle modeling, so that it is neglected in this study.

Other terms from the right member of Equations (1) and (2) have the following expressions:

$$a = \sqrt{3\gamma}, \; c = \sqrt{3RT}, \tag{3}$$

with γ—ratio of specific heat at constant pressure and constant volume, and R—gas specific constant.

The pressure losses due to friction expressed as function of rotation per minute and based on their experimental evaluation for classical thermal engines operating upon Otto and Diesel cycles [81] were adapted to speed [76], and their expression resulted as:

$$\Delta p_f = (0.97 + 0.045w)/N \tag{4}$$

where N—parameter depending on structural characteristics of the engine.

Note that Equations (1) and (2) completed by Equations (3) and (4) clearly show that the finite speed of the piston is responsible for all irreversibility causes, since it appears in both terms in the parentheses.

2.2.2. Model of Carnot Cycle Engine with Analytically Modeled Internal and External Irreversibility

The cyclic system of a Carnot heat engine, including irreversibilities of finite-rate heat transfer between the gas in the thermal engine and its heat reservoirs, heat leakage between the reservoirs, and internal dissipations of the working fluid, is shown schematically in Figure 1 [48,49]. The working fluid in the system is alternately connected to a hot reservoir at constant temperature $T_{H,S}$ and to a cold reservoir at constant temperature $T_{L,S}$ and its temperatures are, respectively, T_H and T_L.

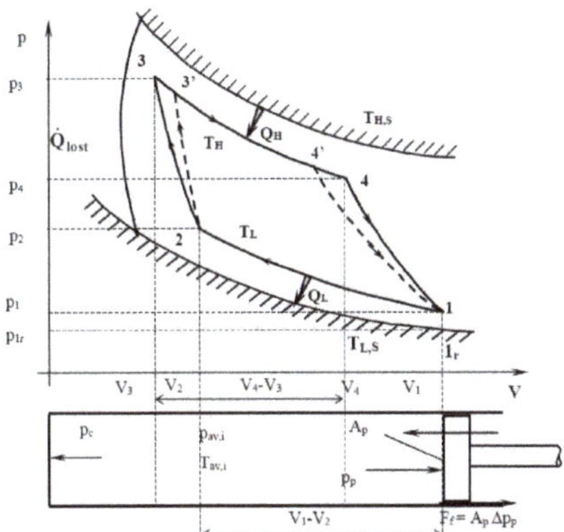

Figure 1. Carnot engine cycle with finite speed of the piston illustrated in p-V diagram [48,49].

Heat losses between the two heat reservoirs temperature level through the engine are considered by the heat rate term $\dot{Q}_{lost}$. In addition, irreversible adiabatic processes are shown by the curves 2-3$'$ and 4$'$-1.

Inside the cylinder with the piston illustrated in the bottom side of Figure 1 appears several pressures that are used in a process with finite speed analysis: on the piston, p_p, on the cylinder, p_c, and the instantaneous average pressure in the gas, $p_{av,i}$.

By integrating Equations (1) and (2) over the isothermal processes of the Carnot cycle, the following expressions for the energy exchanges are dependent of the average piston speed yield:

- The irreversible heat received by the cycle gas from the source:

$$Q_H = z'_H \cdot mRT_H ln\frac{V_4}{V_3} = z'_H \cdot mRT_H \cdot ln\varepsilon,$$ (5)

with z'_H—irreversible coefficient that accounts for a limited heat input in the cycle due to the finite speed of the process:

$$z'_H = \left(1 - \frac{aw}{\sqrt{3RT_H}} - \frac{f \cdot \Delta p_f}{p_{av,34}}\right).$$ (6)

This irreversible coefficient shows that regardless of the heat available at the source, the cycle gas can only receive a limited amount of heat from the source.

- The irreversible heat rejected by the cycle gas to the sink:

$$Q_L = z'_L \cdot mRT_L ln\frac{V_2}{V_1} = -z'_L \cdot mRT_L \cdot ln\varepsilon,$$ (7)

with z'_L—irreversible coefficient that accounts for a limited heat rejected by the cycle gas to the sink due to the finite speed of the process:

$$z'_L = \left(1 + \frac{aw}{\sqrt{3RT_L}} + \frac{f \cdot \Delta p_f}{p_{av,12}}\right).$$ (8)

- The irreversible work produced/consumed during the isothermal processes of the cycle:

$$W_{H,w} = z_H \cdot mRT_H \cdot ln\varepsilon,$$ (9)

$$|W_{L,w}| = z_L \cdot mRT_L \cdot ln\varepsilon,$$ (10)

with the corresponding irreversible coefficients:

$$z_H = \left(1 - \frac{aw}{\sqrt{3RT_H}} - \frac{\Delta p_f}{p_{av,34}}\right),$$ (11)

$$z_L = \left(1 + \frac{aw}{\sqrt{3RT_L}} + \frac{\Delta p_f}{p_{av,12}}\right).$$ (12)

with

$$mR = P_{1r}V_{1r}/T_{1r},$$ (13)

and

$$T_{1r} = T_{L,S}, \; V_{1r} = V_1.$$ (14)

and

$$\frac{V_4}{V_3} = \frac{V_1}{V_2} = \varepsilon.$$ (15)

The work per cycle results from Equations (9) and (10) as:

$$W_{cycle,w} = mR(z_H T_H - z_L T_L)ln\varepsilon.$$ (16)

The non adiabaticity of the engine suggested in Figure 1 by the term $\dot{Q}_{lost}$ is better explained in Figure 2 by the insulating wall between the two semi-cylinders that form the heat conduction path between the heat source and sink.

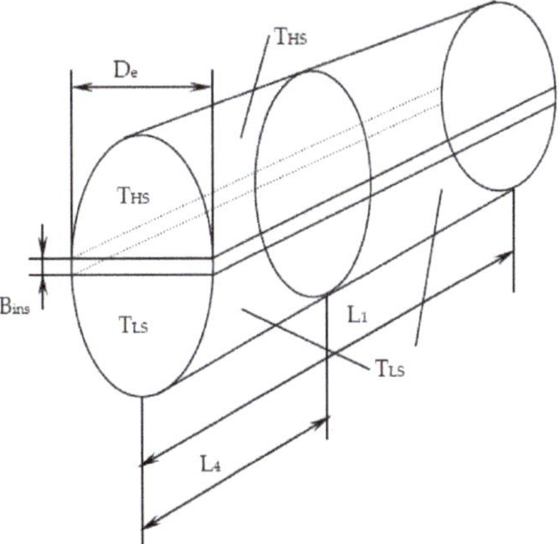

Figure 2. The cylinder configuration used in heat transfer area computation [48,49].

The heat transfer rate lost through this conduction path is:

$$\dot{Q}_{lost} = k_{ins} A_{lost} (T_{HS} - T_{LS}) / B_{ins}, \tag{17}$$

where k_{ins}—thermal conductivity of the insulation, and B_{ins}—insulation thickness.
Equation (17) expressed on the cycle becomes:

$$Q_{lost,cycle} = \dot{Q}_{lost} \cdot \tau_{cycle}. \tag{18}$$

The cycle time duration can be expressed as:

$$\tau_{cycle} = \frac{2(V_1 - V_3)}{w A_p}, \tag{19}$$

with A_p—piston area.

The area associated to the heat transfer rate lost between the source and sink yields (see Figure 2):

$$A_{lost} = (D + 2L_4)(D_e - D), \tag{20}$$

where D is the inner diameter of the cylinder.

This heat transfer rate lost per cycle will modify the heat supply from the source and the heat rejected to the sink as follows:

$$Q_{H,tot} = Q_H + Q_{lost,cycle}, \tag{21}$$

$$|Q_{L,tot}| = |Q_L| + Q_{lost,cycle}. \tag{22}$$

In the above equations, the heat input to the cycle gas and heat rejected from the gas to the sink may be considered those already given by Equations (5) and (7), or it can be expressed in terms of heat transfer as follows:

$$Q_H = U_H(w) \cdot A_H \cdot (T_{H,S} - T_H) \cdot \tau_H, \tag{23}$$

$$|Q_L| = U_L(w) \cdot A_L \cdot (T_L - T_{L,S}) \cdot \tau_L. \tag{24}$$

where $U_H(w)$ and $U_L(w)$ are the overall heat transfer coefficient during the heat exchange at the source and sink, respectively, and A_H and A_L are the area of the heat transfer surfaces.

The heat transfer expressed using the Finite Speed analysis (Equations (5) and (7)) should be the same as the heat transfer corresponding to the above Equations (23) and (24).

Therefore, the two equalities allow expressing the *temperature of the gas at the hot end and at the cold end* respectively, in connection with the source and sink temperature:

$$T_H = T_{H,S} \cdot \left[1 + \frac{z'_H \cdot mR \cdot \ln\varepsilon}{U_H(w) \cdot A_H \cdot \tau_H} \right]^{-1} \tag{25}$$

$$T_L = T_{L,S} \cdot \left[1 - \frac{z'_L \cdot mR \cdot \ln\varepsilon}{U_L(w) \cdot A_L \cdot \tau_L} \right]^{-1}. \tag{26}$$

The overall heat transfer coefficients of the heat exchanger at source and sink, U_L, U_H are calculated based on average bulk fluid temperatures by using well-known equations [82]:

$$Nu_D = \begin{cases} 1.86(Re_D Pr)^{\frac{1}{3}} \left(\frac{D}{L} \right)^{\frac{1}{3}} \left(\frac{\mu}{\mu_{wall}} \right)^{0.14}, & for\ Re_D \leq 2300 \\ 0.023\ Re_D^{0.8} Pr^n, & for\ Re_D \geq 3000 \end{cases}, \tag{27}$$

with $n = 0.4$ for heating, respectively, $n = 0.3$ for cooling.

Similarly, the dynamic viscosity and the thermal conductivity of the gas are calculated using polynomial functions [64], based on the bulk gas temperature.

The contact time per cycle for the heat transfer from the heat source to the engine corresponding to the isothermal process is:

$$\tau_H = (L_4 - L_3)/w = \frac{L_1 \left(1 - \frac{1}{\varepsilon} \right) \left(\frac{T_L}{T_H} \right)^{\frac{1}{\gamma}-1}}{w}, \tag{28}$$

while the contact time per cycle for heat transfer from the gas engine to the sink is:

$$\tau_L = (L_1 - L_2)/w = \frac{L_1 \left(1 - \frac{1}{\varepsilon} \right)}{w}. \tag{29}$$

The area for the heat transfer between the source and the hot gas during the isothermal heat addition process (see Figure 2) is:

$$A_H = 0.5D \left(\frac{\pi D}{4} - B_{ins} \right) + 0.5L_1 \left(1 + \frac{1}{\varepsilon} \right) \left(\frac{\pi D}{2} - B_{ins} \right) \cdot \left(\frac{T_L}{T_H} \right)^{\frac{1}{\gamma}-1}. \tag{30}$$

Similarly, the area for heat transfer between the cold gas and the sink during the isothermal heat rejection process is expressed as:

$$A_L = 0.5D \left(\frac{\pi D}{4} - B_{ins} \right) + 0.5L_1 \left(1 + \frac{1}{\varepsilon} \right) \left(\frac{\pi D}{2} - B_{ins} \right), \tag{31}$$

with

$$\frac{L_1}{\varepsilon} = L_2. \tag{32}$$

The power output of the irreversible Carnot engine is given by:

$$P_{\Delta T,w,Q_{lost}} = \frac{W_{cycle,w}}{\tau_{cycle}}. \tag{33}$$

The efficiency of the Carnot cycle with internal and external irreversibility is:

$$\eta_{\Delta T,w,Q_{lost}} = 1 - \frac{|Q_{L,w}|}{Q_{H,w}} = 1 - \frac{T_L}{T_H} \cdot \frac{z'_L}{z'_H}. \tag{34}$$

Then, the entropy generation per cycle can be expressed as:

$$\Delta S_{cycle} = \frac{Q_{H,w}}{T_H} + \frac{Q_{L,w}}{T_L} = mRln\varepsilon \cdot \left(z'_H - z'_L\right),\tag{35}$$

and its corresponding expression per unit time is:

$$\dot{S}_{gen} = \frac{\Delta S_{cycle}}{\tau_{cycle}}.\tag{36}$$

The results of this optimization model will be given in Section 3.

2.3. The Curzon–Ahlborn Model of the Carnot Cycle Engine Combined with the Analysis Based on Thermodynamics with Finite Speed (TFS)

The model aims to combine the analysis of the Carnot cycle engine with only external irreversibility in Thermodynamics in Finite Time (FTT) with the main advantage of the Thermodynamics with Finite Speed (TFS) approach, namely the internal irreversibility quantification as a function of the speed of the process.

The main differences of this model compared to the previous one are represented by:

- The absence of heat losses Q_{lost}, in order to consider similar cycles in both analyses.
- The presence of losses in the work expression, so that the work lost in the two adiabatic processes due to finite speed is obtained by integrating the irreversible work for processes with finite speed in the processes 2-3′ and 4′-1 (Equation (2)) and subtracting the reversible work in the processes 2-3 and 4-1 (see Figure 1):

$$W_{lost,\,ad,\,int} = \left(\frac{aw}{c_{23'}} + \frac{\Delta p_f}{p_{23'}}\right)(V_{3'} - V_2)_{23'} - \left(\frac{aw}{c_{4'1}} + \frac{\Delta p_f}{p_{4'1}}\right)(V_1 - V_{4'})_{4'1}.\tag{37}$$

where $p_{23'}$ and $p_{4'1}$ are the average gas pressure on the irreversible adiabatic compression and expansion, respectively.

This lost work term is then subtracted from the work per cycle given by Equation (16), since it does not include the effect of internal irreversibilities of the adiabatic processes.

By including this lost work term in the analysis, an expression for the efficiency of the Carnot cycle, considering all internal and external irreversibilities yields as:

$$\eta_{\Delta T,w,f} = \left(\frac{z_H}{z'_H} - \frac{z_L \cdot T_L}{z'_H \cdot T_H}\right) - I_{ad}\frac{1 - T_L/T_H}{z'_H(\gamma - 1)ln\varepsilon'},\tag{38}$$

where the irreversible adiabatic process contribution of the internal irreversibility of the cycle, due to the finite piston speed and friction, I_{ad}, results as:

$$I_{ad} = aw\left(\frac{1}{c_{23'}} + \frac{1}{c_{4'1}}\right) + \Delta p_f\left(\frac{1}{p_{23'}} + \frac{1}{p_{4'1}}\right).\tag{39}$$

Note that the second term in Equation (38) is obtained by integration of the First Law for Processes with Finite Speed (TFS) for the adiabatic processes 23′ and 4′1 (see Figure 1), Equations (1) and (2).

The combination of the two analyses based on FTT and TFS models will include a similar term to that given by Equation (39) in the Curzon–Ahlborn approach. As previously mentioned, this approach included the time duration of the cycle processes, with the assumption that the adiabatic processes occur rapidly and accordingly consume far less time than the isothermal processes. Based on this assumption, the FTT and TFS analyses can be rationally compared only if the Carnot cycle engine dimensions and number of cycles per unit time are made equal in both cases. In a TFS analysis, the speed of the piston, w, is assumed constant in each of the four processes and equals the average speed based on the number of cycles per unit time. However, in a Curzon–Ahlborn type analysis (FTT optimization), the speed of isothermal compression w_L, the speed of isothermal expansion w_H, and the speed of the adiabatic processes w_{ad} (assumed equal for both adiabatic processes), are calculated. The result must be consistent with the total cycle time optimized for maximum power.

When this comparison is performed, the following process speeds, in terms of the average speed, are obtained (see Figure 2) [49]:

$$w_L = \frac{a'(L_1 - L_2)(1 + Z^*)}{2L_1/w}, \tag{40}$$

$$w_H = \frac{a'(L_4 - L_3)(1/Z^* + 1)}{2L_1/w}, \tag{41}$$

$$w_{ad} = \frac{a'w[(L_2 - L_3) + (L_1 - L_4)]}{2L_1(a' - 1)}, \tag{42}$$

where Z^*—ratio of the optimized duration of the isothermal processes in the Curzon–Ahlborn treatment (FTT), a'—coefficient depending on time to speed transfer.

The optimized temperatures in the Curzon–Ahlborn analysis [3] are expressed based on corresponding optimized times for each process, as follows:

$$T_{L,FTT} = T_L \frac{1 + \sqrt{\frac{T_H}{T_L} \cdot \frac{1}{Z^*}}}{1 + \frac{1}{Z^*}}, \tag{43}$$

$$T_{H,FTT} = T_H \frac{1 + \sqrt{\frac{T_L}{T_H} \cdot Z^*}}{1 + Z^*}. \tag{44}$$

By using the above expressions of temperatures and including the effect of internal irreversibility, the corresponding power of Carnot cycle in FTT analysis is:

$$Power_{FTT} = \frac{\frac{A_L U_L}{a'} \cdot \left(\sqrt{T_H} - \sqrt{T_L}\right)^2}{(Z^* + 1)^2} - (W_{loss,ad,int} + W_{loss,isot,int}) \frac{1}{\tau_{cycle}}. \tag{45}$$

Equation (45) appears as a combination of the two analyses as the first term is the original Curzon–Ahlborn term [3] taking account of only external irreversibilities generated by the temperature difference, and the second term accounts for internal irreversibilities generated by the finite speed and friction from the TFS approach.

Nevertheless, a simpler expression of the power output can be also given as:

$$Power_{\Delta T,w,f,FTT} = Q_H \cdot \eta'_{\Delta T,w,f,FTT} \cdot \frac{1}{\tau_{cycle}}, \tag{46}$$

where the efficiency term contains all irreversibility causes of the Carnot cycle engine.

The passage from the efficiency of the Carnot cycle including only external irreversibilities and corresponding to maximum power output in the original Curzon–Ahlborn analysis [3]:

$$\eta_{\Delta T,FTT} = 1 - \frac{T_{L,FTT}}{T_{H,FTT}} = 1 - \sqrt{\frac{T_{LS}}{T_{HS}}}, \tag{47}$$

will be performed here by including the effects of internal irreversibilities. Similarly, Equations (5)–(12) are expressed by evaluating Z_{FTT} and Z'_{FTT} irreversible coefficients at the appropriate speeds (w_L and w_H) on the isothermal processes at T_L and T_H respectively, and on the adiabatic processes (w_{ad}) conveying to the following corrected efficiency:

$$\eta_{irr,int,FTT} = \frac{Z_{H,FTT}}{Z'_{H,FTT}} - \frac{Z_{L,FTT} \cdot T_{L,FTT}}{Z'_{L,FTT} \cdot T_{H,FTT}} - I'_{ad} \frac{1 - T_{L,FTT}/\cdot T_{H,FTT}}{Z'_{H,FTT}(\gamma - 1)\ln\varepsilon} \tag{48}$$

where the equivalent term I_{ad}' to that from Equation (39) is similar, but it is based on w_{ad} (Equation (42)) instead of w and also on the resulting temperatures and pressures from the Curzon–Alhborn. Ref. [3] analysis of the Carnot cycle completed by TFS tools (Equations (43) and (44)).

2.4. Unification Attempts of Thermodynamics in Finite Time and Thermodynamics with Finite Speed Analyses

The first unification attempt is based on [47] that had a very important role in the development of Thermodynamics with Finite Speed (TFS) and the Direct Method, for analytical evaluation of the performances of irreversible cycles with internal and external irreversibilities. Later, it was completed by [31,34].

Specific issues addressed in this model are illustrated on cycle Carnot engine represented in *T-S* coordinates in Figure 3. There are shown to have external irreversibility due to heat transfer from the source (with fixed temperature $T_{H,S}$) to the cycle temperature at the hot end, T_X, during the isothermal heat addition process 2–3. Then, internal irreversibilities due to the finite piston speed are considered during only the adiabatic compression and expansion processes. The sink temperature and the cycle temperature at the cold end are the same. The sink temperature, T_0, is fixed, while the cycle temperature at the hot end, T_X, is a variable.

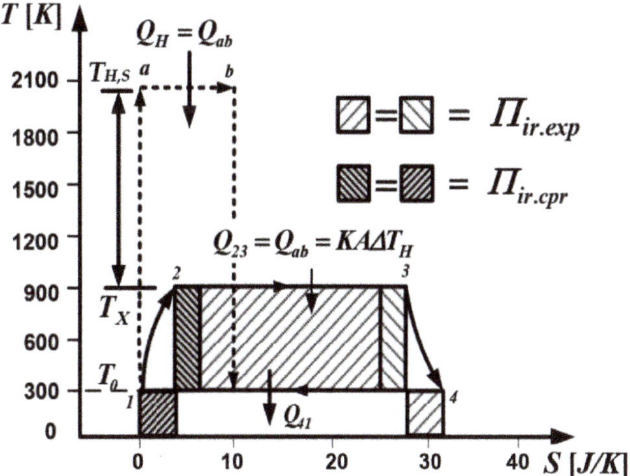

Figure 3. Carnot engine cycle with internal irreversibilities illustrated in T-S diagram [47,52].

Another novelty compared to previous model consists of the use of entropy variation calculation on the irreversible cycle processes that will provide a term in the cycle efficiency expression that could unify the two analyses.

The first unification attempt is based on the First Law of Thermodynamics for Processes with Finite Speed [70–73] in its reduced form that considers only the internal irreversibility due to the finite speed of the piston:

$$dU = \delta Q - p_{av,i}\left(1 \pm \frac{aw}{c}\right)dV. \tag{49}$$

From the equation for adiabatic irreversible processes of ideal gases with constant specific heats that is derived from Equation (49) by integration [72,73,75,76], one can express the temperature T_2 at the end of an irreversible adiabatic process as

$$T_2 = \frac{\left(1 \pm \frac{aw}{c_1}\right)^2}{\left(1 \pm \frac{aw}{c_2}\right)^2} T_1 \left(\frac{V_1}{V_2}\right)^{\gamma-1} = \delta_{irr} T_1 \left(\frac{V_1}{V_2}\right)^{\gamma-1}, \tag{50}$$

where γ is the ratio of the specific heat at constant pressure and at constant volume.

For a compression process with finite speed $w \ll c$, one could express $\delta_{irr.cpr}$ as follows:

$$\delta_{irr,cpr} = \frac{\left(1 + \frac{aw}{c_1}\right)^2}{\left(1 + \frac{aw}{c_2}\right)^2} \cong \left[\left(1 + \frac{aw}{c_1}\right)\left(1 - \frac{aw}{c_2}\right)\right]^2 = \left[1 + \frac{aw}{c_1} - \frac{aw}{c_2}\right]^2, \tag{51}$$

if $a^2 w^2 \ll c_1 \cdot c_2$ and the corresponding term is neglected.

Note that for compression, the plus sign is used in parenthesis.

Note that the average molecular speed c_2 depends on temperature T_2 that contains $\delta_{irr.cpr}$. Thus, the calculation should be done by using approximations.

The first approximation considers the temperature at the end of the reversible adiabatic compression for which one gets (see Equation (3)):

$$T_2 = T_1 \left(\frac{V_1}{V_2}\right)^{\gamma-1} \Rightarrow c_2 = c_1 \left(\frac{V_1}{V_2}\right)^{\frac{\gamma-1}{2}}. \tag{52}$$

By substituting Equation (52) in Equation (51), a first evaluation of $\delta_{irr.cpr}$ is done:

$$\delta_{irr.cpr} = \left[1 + \frac{aw}{c_1} - \frac{aw}{c_1}\left(\frac{V_2}{V_1}\right)^{\frac{\gamma-1}{2}}\right]^2. \tag{53}$$

Note that a more precise approximation is possible by combining Equations (50) and (53) that yields:

$$T_2 = \delta_{irr.cpr} T_1 \left(\frac{V_1}{V_2}\right)^{\gamma-1}, \tag{54}$$

and a better approximation for the adiabatic irreversible coefficient is given by:

$$\delta'_{irr.cpr} = \left[1 + \frac{aw}{c_1} - \frac{aw}{c_1}\left(\frac{V_2}{V_1}\right)^{\frac{\gamma-1}{2}}(\delta_{irr,cpr})^{-\frac{1}{2}}\right]^2. \tag{55}$$

For simplicity, the first approximation expression of the adiabatic irreversible coefficient (Equation (53)) is used hereafter.

The entropy variation computation in the case of an adiabatic irreversible process of compression with finite speed when the results from Equations (50) and (53) are introduced in the classical formula of ΔS:

$$\Delta S = S_f - S_i = mc_v ln\frac{T_f}{T_i} + mRln\frac{V_f}{V_i}, \tag{56}$$

which provides:

$$\Delta S_{irr,cpr} = mc_v ln \left[1 + \frac{aw}{c_1} - \frac{aw}{c_1}\left(\frac{V_2}{V_1}\right)^{\frac{\gamma-1}{2}}\right]^2. \tag{57}$$

Similarly, the entropy variation expression on the adiabatic irreversible expansion can be derived showing that the only difference consists in the change of signs in the parentheses, so that one can give a general form of both compression and expansion processes, as:

$$\Delta S^w_{ad,irr} = mc_v ln \left[1 \pm \frac{aw}{c_1} \mp \frac{aw}{c_1}\left(\frac{V_2}{V_1}\right)^{\frac{\gamma-1}{2}} \right]^2 . \tag{58}$$

By using Equations (56) and (58) in the present analysis on the two irreversible adiabatic processes and on the isothermal expansion, the following expressions result:

$$\Delta S^w_{ad.irr.cpr} = \Delta S_{12} = mc_v ln(\alpha_1), \text{ with } \alpha_1 = \left[1 + \frac{aw_{cpr}}{c_1} - \frac{aw_{cpr}}{c_1}\left(\frac{V_2}{V_1}\right)^{\frac{\gamma-1}{2}} \right]^2 , \tag{59}$$

$$\Delta S^w_{ad.irr.exp} = \Delta S_{34} = mc_v ln(\alpha_2), \text{ with } \alpha_2 = \left[1 - \frac{aw_{exp}}{c_3} + \frac{aw_{exp}}{c_3}\left(\frac{V_4}{V_3}\right)^{\frac{\gamma-1}{2}} \right]^2 , \tag{60}$$

$$\Delta S_{23} = S_3 - S_2 = mRln\frac{p_2}{p_3}. \tag{61}$$

with c_v—specific heat at constant volume, R—specific constant of the cycle fluid.

Then, the actual thermal efficiency of the Carnot cycle engine with irreversibilities can be expressed based on previous calculation (see Figure 3) as:

$$\eta_{act} = 1 - \frac{Q_{41}}{Q_{23}} = 1 - \frac{T_C \Delta S_{14}}{T_X \Delta S_{23}} = 1 - \frac{T_0(\Delta S_{23} + \Delta S_{12} + \Delta S_{34})}{T_X \Delta S_{23}}, \tag{62}$$

and together with Equations (59)–(61), the following expression results:

$$\eta_{act} = 1 - \frac{T_0}{T_X}\left[1 + \frac{2ln(\alpha_1\alpha_2)}{(\gamma-1)ln\frac{p_2}{p_3}} \right]. \tag{63}$$

When the piston speed is much less than the average molecular speed, namely $aw_{cpr} \ll c_1$, and $a_{exp} \ll c_3$, one gets a simplified form of Equation (63):

$$\eta_{act} = 1 - \frac{T_0}{T_X}\left[1 + \frac{2(\beta_1 + \beta_2)}{(\gamma-1)ln\frac{p_2}{p_3}} \right], \tag{64}$$

where

$$\beta_1 = \frac{aw_{cpr}}{c_1}\left(1 - \sqrt{\frac{T_0}{T_X}} \right), \tag{65}$$

$$\beta_2 = \frac{aw_{exp}}{c_3}\left(\sqrt{\frac{T_X}{T_0}} - 1 \right). \tag{66}$$

For the same speed of the piston on the two adiabatic processes of the cycle, Equation (64) becomes:

$$\eta_{act} = 1 - \frac{T_0}{T_X}\left\{ 1 + \frac{4aw}{c_1}\frac{\left(1 - \sqrt{\frac{T_0}{T_X}}\right)}{(\gamma-1)ln\frac{p_2}{p_3}} \right\}. \tag{67}$$

Once having the actual efficiency of the cycle, the power output of the engine can be easily derived as:

$$\dot{W}_{act} = \dot{Q}_H \eta_{act} = U_H A_H(T_{H,S} - T_X)\eta_{act}. \tag{68}$$

To render the model more general, a non-dimensional form of the power output of the Carnot engine will be optimized, namely:

$$P_{ND} = \frac{\dot{W}_{act}}{U_H A_H T_{H,S}}.$$
(69)

Moreover, the actual efficiency is expressed as a product of the Carnot reversible efficiency:

$$\eta_{CC} = \left(1 - \frac{T_0}{T_X}\right),$$
(70)

and the second law efficiency accounting for irreversibilities:

$$\eta_{IIad.irr}^{w} = \left[1 - \frac{C\left(\frac{T_0}{T_X}\right)}{\left(1 + \sqrt{\frac{T_0}{T_X}}\right)}\right],$$
(71)

with the internal irreversible coefficient C given by:

$$C = \frac{4aw}{c_1(\gamma - 1)\ln\frac{p_2}{p_3}}.$$
(72)

By combining Equation (69) with Equations (68), (70)–(72) and term rearrangement, one gets:

$$P_{ND} = \left(1 - \frac{T_X}{T_{H,S}}\right)\left(1 - \frac{T_0}{T_X \Phi}\right),$$
(73)

With

$$\Phi = \frac{1}{1 + C\left(1 - \sqrt{\frac{T_0}{T_X}}\right)}.$$
(74)

Note that for a given cycle fluid, coefficient Φ depends only on the fluid temperature at the hot end, T_X, and the piston speed, w. Thus, the non-dimensional power (Equation (73)) is seen to be a complex function of T_X and the piston speed by the term C. Searching for an analytic expression of the optimum temperature to maximize the non-dimensional power can be done in the first approximation, for Φ = constant in Equation (73). This is in good agreement with Ibrahim's approach [16], where for Φ constant, the expression of the optimal temperature of the cycle fluid at the hot end that maximizes the power output of the engine was established as:

$$T_X^{\max P_{ND}} \rightarrow T_{opt} = \sqrt{\frac{T_{H,S} \cdot T_0}{\Phi}}.$$
(75)

Although this is a simple expression, the value of Φ is not known. It is indicated as a parameter with a given (not computed) value.

In the present analysis, one can approximate the value of T_{opt} by iterations. Thus:

- For $w = 0$, which means an internally reversible cycle, Equations (72) and (74) lead to $\Phi = 1$, so that Equation (75) becomes:

$$T_{opt}^{(w=0)} = \sqrt{T_{H,S} \cdot T_0}.$$
(76)

- For $w \neq 0$, by combining Equations (74) and (76), a first approximation of the term responsible for cycle irreversibilities is expressed as:

$$\Phi_w = \left[1 + C\left(1 - \sqrt[4]{\frac{T_0}{T_{H,S}}}\right)\right]^{-1},$$
(77)

and the corresponding optimum temperature yields from Equation (75) as:

$$T_{opt}^{(w\neq 0)} = \sqrt{T_{H,S} \cdot T_0 \left[1 + C\left(1 - \sqrt[4]{\frac{T_0}{T_{H,S}}}\right)\right]}. \tag{78}$$

Equation (78) is the first approximation of the optimum temperature to maximize the non-dimensional power when the piston speed is not zero and when therefore both internal and external irreversibilities are accounted for.

Furthermore, the next step in the approximation procedure is to replace T_x in Equation (74) by Equation (78), that allows obtaining a more accurate expression of Φ term:

$$\Phi'_w = \left[1 + C\left(1 - \sqrt[4]{\frac{T_0 \Phi_w}{T_{H,S}}}\right)\right]^{-1}. \tag{79}$$

One could continue the iteration, but the gain in accuracy would become insignificant. Thus, the optimized temperature of the cycle fluid at the hot end of the engine coming out of TFS analysis is:

$$T_{opt}^{'(w\neq 0)} = \sqrt{\frac{T_{H,S} \cdot T_0}{\Phi'_w}}, \tag{80}$$

and the maximum non dimensional power output of the internally and externally irreversible Carnot cycle becomes:

$$P_{ND,maxZ} = \left(1 - \frac{T_{opt}^{'(w\neq 0)}}{T_{H,S}}\right)\left(1 - \frac{T_0}{\Phi'_w T_{opt}^{'(w\neq 0)}}\right) = \left(1 - \sqrt{\frac{T_0}{T_{H,S}\Phi'_w}}\right)^2. \tag{81}$$

Then, the efficiency of the irreversible Carnot cycle is calculated by substituting $T_{opt}^{'(w\neq 0)}$ into Equation (67) that leads to:

$$\eta_{act} = 1 - \sqrt{\frac{T_0}{T_{H,S}}\Phi'_w} \cdot \left[1 + C\left(1 - \sqrt[4]{\frac{T_0}{T_{H,S}}\Phi'_w}\right)\right]. \tag{82}$$

One can see now that Equation (82) unifies the FTT and TFS analyses by the same expression of the actual efficiency of an irreversible Carnot cycle engine. Thus:

- For internally reversible, externally irreversible Carnot cycle engine for which $w = 0$ and consequently, $\Phi'_w = 1$, one gets the Curzon–Ahlborn "nice radical" [3]:

$$\eta_{CA} = 1 - \sqrt{\frac{T_0}{T_{H,S}}}. \tag{83}$$

- For an internally and externally irreversible Carnot cycle engine for which $w \neq 0$ and consequently, $\Phi'_w > 1$, one gets:

$$\eta_{act} = 1 - \sqrt{\frac{T_0}{T_{H,S}}}\zeta_w, \tag{84}$$

with

$$\zeta_w = \sqrt{\Phi'_w}\left[1 + C\left(1 - \sqrt[4]{\frac{T_0}{T_{H,S}}\Phi'_w}\right)\right]. \tag{85}$$

Note that $\zeta_w \geq 1$ and it accounts for internal irreversibilities of the cycle when depending on the piston speed. Equations (83)–(85) clearly show that the nice radical of FTT analysis overestimates the actual efficiency of the engine evaluated by TFS analysis.

A second unification attempt is under development. It aims to extend the modeling by considering, in addition to the finite speed, two other causes of internal irreversibility: friction and throttling.

Based on previous equations of the first unification attempt, a new expression was derived for the actual efficiency of the Carnot cycle engine:

$$\eta_{act}^{irr} = 1 - \frac{T_0}{T_X}\left\{1 + 4\left(\frac{aw}{c_1} + \frac{\Delta p_f}{p_{av,34}} + \frac{\Delta p_{thr}}{p_{av,34}}\right)\frac{\left(1 - \sqrt{\frac{T_0}{T_X}}\right)}{(\gamma - 1)ln\frac{p_2}{p_3}}\right\},\tag{86}$$

where Δp_{thr} is estimated as [62–64,83]:

$$\Delta p_{thr} = C_{thr} \cdot w^2,\tag{87}$$

with $C_{thr} = 0.005$.

Then, the irreversibility coefficient yields:

$$C_{irr} = 4\left(\frac{aw}{c_1} + \frac{\Delta p_f}{p_{av,34}} + \frac{\Delta p_{thr}}{p_{av,34}}\right)\frac{1}{(\gamma - 1)ln\frac{p_2}{p_3}}.\tag{88}$$

The power output and efficiency of the Carnot cycle engine with finite speed processes considering all internal irreversibility causes are smaller compared to those determined from Equations (81) and (82), since the new correction is more substantial by its three terms (Equation (88)).

The results of this modeling emphasize optimum speed values generating maximum power output, as well as the effect of irreversibilities on the optimum cycle high temperature.

3. Results

The results of TFS analysis presented in Section 2.2 relative to a Carnot cycle engine with internal and external irreversibilities generated by losses due to (1) heat transfer between the cycle and the heat source and sink, (2) the effect of variation in the area for heat transfer and in the dwell time for heat transfer due to the movement of the piston during the isothermal expansion and compression processes, and (3) non adiabaticity of the engine are presented in Figures 4–6. The following fixed parameters entering in the equations of the model were used: $D = 0.015$ m; $L_1 = 2$ m; $\varepsilon = 3$; $f = 0$; $p_{1r} = 0.05$ bar (pressure of the gas in state 1r); $\Delta p_f = (0.97 + 0.045 \text{ w})/80$; $T_{H,S} = 1200$ K; $T_{L,S} = 300$ K; $\gamma = 1.4$; $B_{ins} = 0.002$ m; $k_{ins} = 0.01$ W/mK; $D_e = 0.019$ m. The cycle fluid is air that is considered as an ideal gas with specific heat, conductivity, and viscosity varying as a function of temperature.

Figure 4 illustrates the effect of irreversibilities introduced gradually on the power output showing the important difference between the cycle power output for the reversible Carnot cycle and for the Carnot cycle with irreversiblities due to the finite speed of the piston. Then, the cycle efficiency including internal and external irreversibilities, $\eta_{\Delta T, w, Q_{lost}}$, is represented as a function of piston speed showing optimum values for maximum performance. In addition, the time rate of entropy generation is added in order to compare the optimization results in terms of optimal speed.

One can see that the piston speed for maximum efficiency is only 4 m/s, for which the rate of entropy generation (per unit of time) is very low. Moreover, the piston speed for maximum power is near 17 m/s, and the rate of entropy generation (per unit of time) at this speed is significantly higher. As expected, the power output decreases, as additional irreversibilities are included in the analysis.

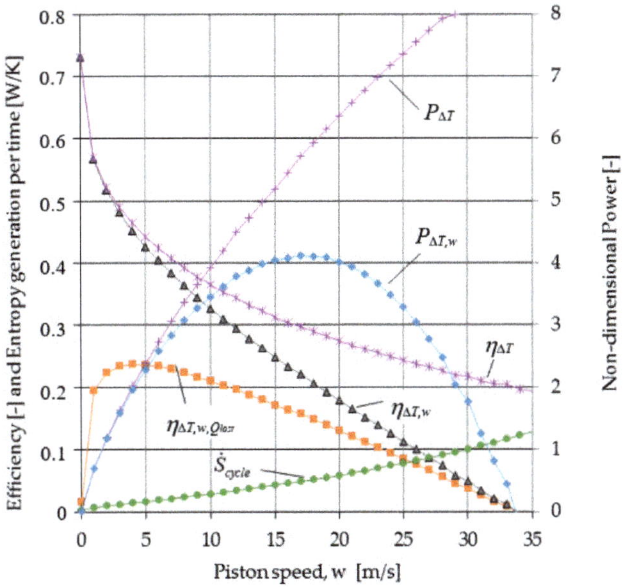

Figure 4. Power, efficiency and entropy generation per time as a function of the piston speed.

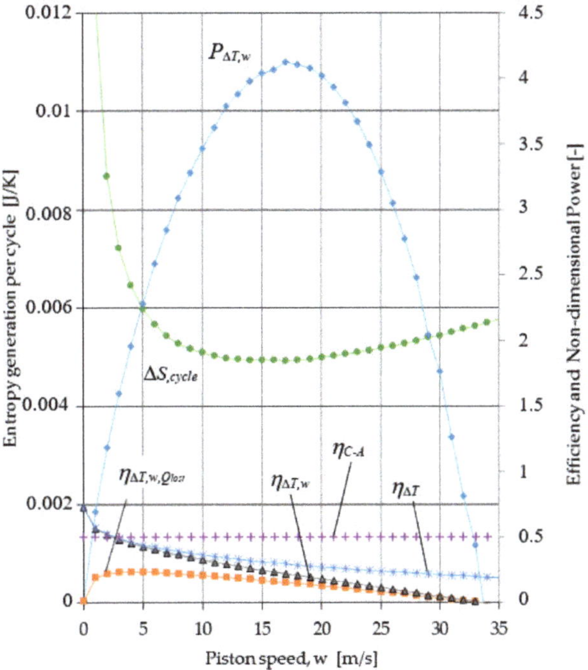

Figure 5. Power, efficiencies and entropy generation per cycle as a function of average piston speed.

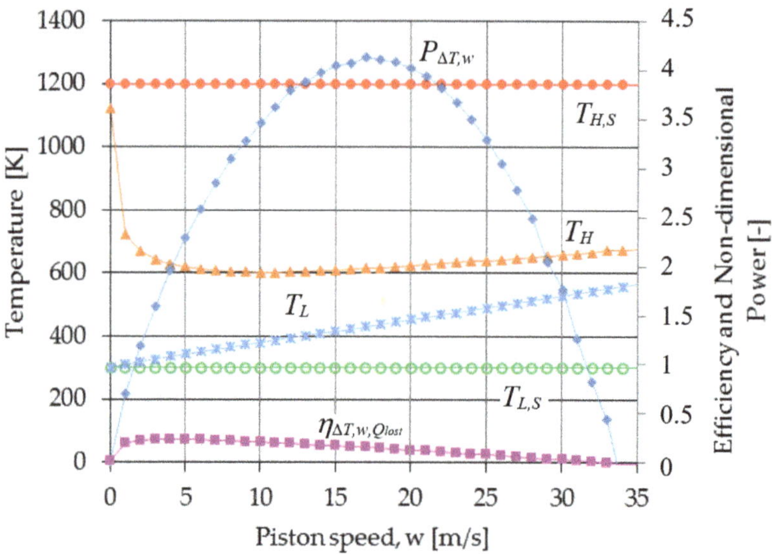

Figure 6. Power, efficiencies and temperatures as a function of average piston speed.

Figure 5 brings together the efficiency of the Carnot cycle determined by the TFS analysis when it is gradually affected by irreversibility, the one based on Curzon–Ahlborn analysis, the power output, and the entropy variation per cycle as functions of piston speed. The efficiency of the Carnot cycle as determined by TFS analysis is at all piston speeds less than the efficiency based on the Curzon–Ahlborn analysis. In addition, for piston speeds greater than w_{opt}, the efficiency of the Carnot cycle at maximum power as determined by TFS is less than the efficiency based on the Curzon–Ahlborn analysis, even if only the external irreversibility is included. For example, the TFS efficiency, at the speed corresponding to maximum power, is 0.29 when only external irreversibilities are included and is 0.15 when both internal and external irreversibilities are included in the analysis.

An important aspect is related to the entropy generation per cycle and per time as functions of piston speed from Figures 4 and 5. Their evolution with the piston speed is completely different, in that only ΔS_{cycle} shows a minimum for the speed as the maximum power output.

The hot and cold heat reservoir temperatures, the hot and cold end gas temperatures, and the Curzon–Ahlborn optimized temperature are shown in Figure 6 as a function of the piston speed. The hot-end gas temperature optimized for maximum power is shown to be nearly the same over a large variation range of piston speeds (5 to 10 m/s), as the Curzon–Ahlborn optimized temperature. In addition, the predicted temperature difference between the high and low gas temperature is shown to increase as the piston speed decreases and to be especially great at piston speeds less than the speed for maximum efficiency.

Some results of the second model (Section 2.3) are shown in Figures 7–9.

Figure 7 illustrates the relative speed of the adiabatic processes and of each of the isothermal processes in FTT optimization compared to the average speed of the piston considered in TFS optimization. The curves show that the optimization results in lower speed than the average speed of the piston w_{TFS}, for the two isothermal processes in FTT optimization. In addition, the high temperature isothermal process has the lowest speed; then, it follows the low temperature isothermal process with a higher speed, while the adiabatic processes occur at a much higher speed. However, the internal irreversibilities were not included in the original Curzon–Ahlborn analysis [3], so the high piston speed during the adiabatic process had no negative effect on the cycle efficiency and power.

In fact, the resulting slower piston speed during the isothermal processes significantly enhanced the cycle efficiency and power in FTT optimization.

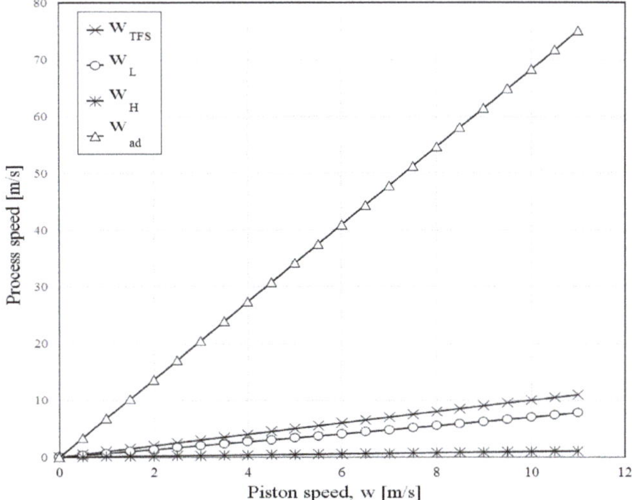

Figure 7. Piston speeds for process in TFS and FTT analyses as function of average piston speed.

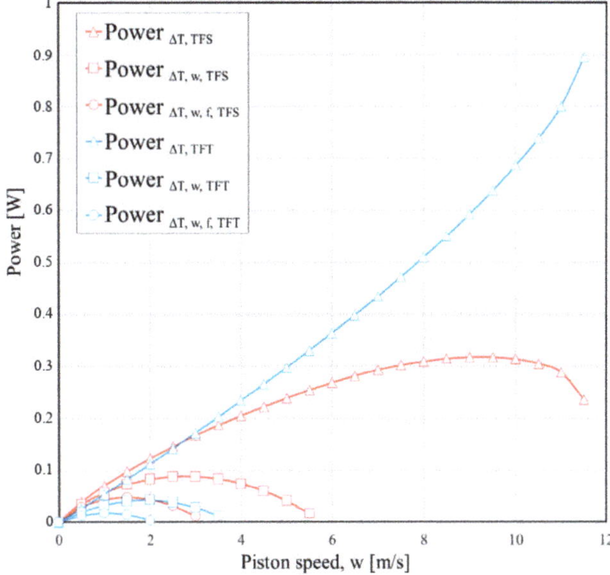

Figure 8. Power output of the Carnot engine for processes in TFS and FTT optimizations as function of average piston.

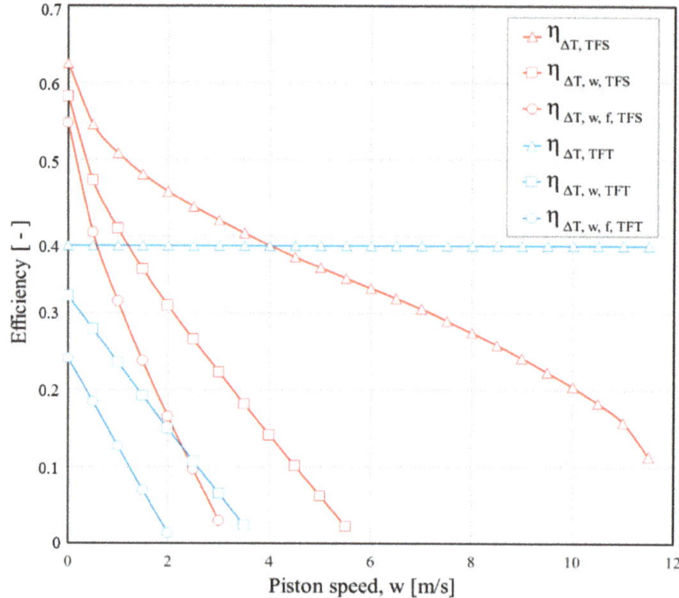

Figure 9. Carnot cycle efficiency based on TFS and FTT optimizations as function of average piston speed.

The effect of the piston speed on the power output and efficiency for a Carnot engine with external irreversibilities and internal ones gradually introduced in both TFS and FTT analyses is shown in Figures 8 and 9, respectively. These results are based on the following fixed parameters: D = 0.015 m; L_1 = 0.5 m; ε = 2; f = 0.3; a' = 1.1; p_{1r} = 0.01 bar (pressure of the gas in state 1_r); Δp_f = (0.97 + 0.045 w)/60 bar; T_{HS} = 800 K; T_{LS} = 300 K; γ = 1.4.

The FTT optimization predicts greater power output from the Carnot engine at almost all piston speeds than the TFS optimization when only external irreversibilities (ΔT) are considered. It is due to the little cycle time that was allocated to the adiabatic processes in the FTT optimization. This allowed more time for the isothermal processes without any penalty associated with the more rapid adiabatic processes, since the internal irreversibilities of these processes are not considered. In the TFS optimization for example, at 9 m/s the power is 0.33 W, and the efficiency is 25%. In the FTT optimization at the same speed, by comparison, the power is 0.6 W, and the efficiency is 39%. However, when the internal irreversibilities are included in the analyses, the TFS optimization results in greater power and efficiency than FTT, even though both are less than when the internal irreversibilities were neglected.

It is also important to keep in mind that a cycle that operates with three different piston speeds for the four processes presents a huge mechanical complication in the design of the actual engine. While it may be possible to design such an engine (for example, using cams with different profiles for each process), there is no need to do so, since the TFS optimization predicts superior operating performance.

The non-dimensional power as determined from Equation (77) as a function of the cycle high temperature and the piston speed is shown in Figure 10. In addition, the power output of the reversible Carnot cycle is added for comparison purposes. The non-dimensional power reveals the maximum value for any fixed piston speed or internal irreversibility consequence, and this maximum is moving toward growing temperature T_x as the piston speed increases.

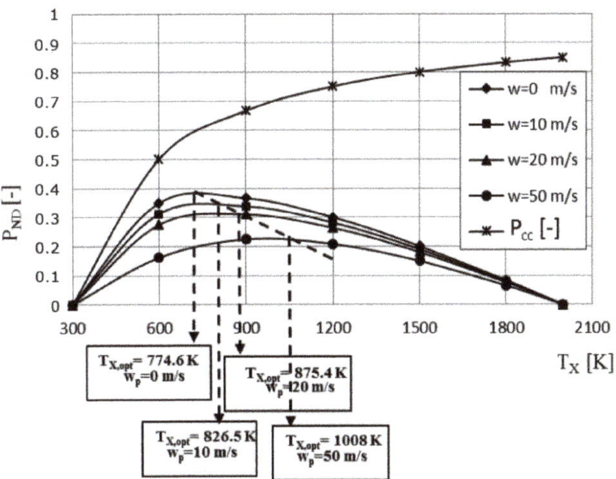

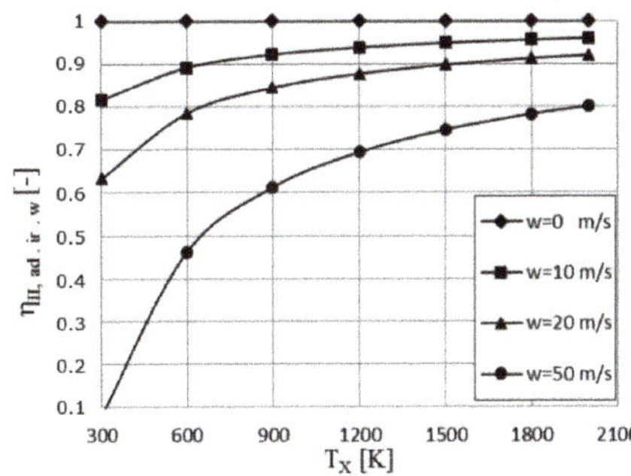

Figure 10. The non-dimensional power of the Carnot cycle engine as a function of the cycle high temperature and the piston speed w, as determined from TFS analysis.

Figure 11 presents the second law efficiency variation versus the cycle high temperature for different values of the piston speed. The curves show that this irreversibility coefficient decreases as piston speed increases, as expected, and the decrease is more important at lower values of the cycle high temperature.

Figure 11. The effect of the piston speed, w, on the second law efficiency variation with the cycle high temperature.

Regarding the irreversible term Φ determined from Equation (74), its variation with the cycle high temperature and piston speed becomes important mainly at high speeds, as illustrated in Figure 12. However, there is little change of Φ in the region of optimal temperatures (from 800 to 1000 K).

The comparison of the results before (Figure 10) and after (Figure 13) using approximations in search of optimal temperature expression that optimizes the power output of the engine shows good agreement and lends confidence that a first iteration provides sufficiently accurate results for most purposes. However, it is possible to improve the accuracy of the results by making a new iteration.

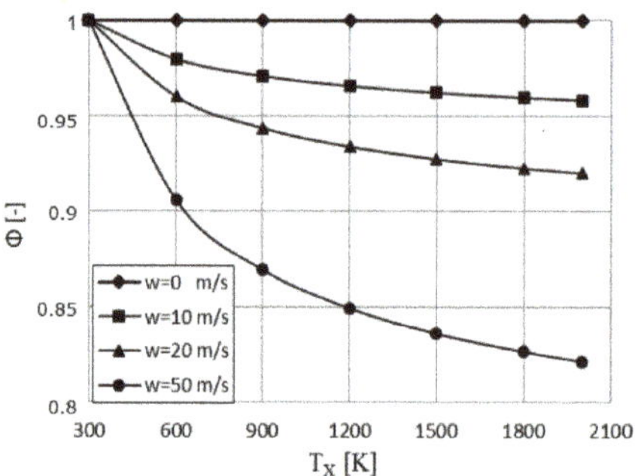

Figure 12. Parameter Φ variation with the cycle high temperature T_X for different piston speed values.

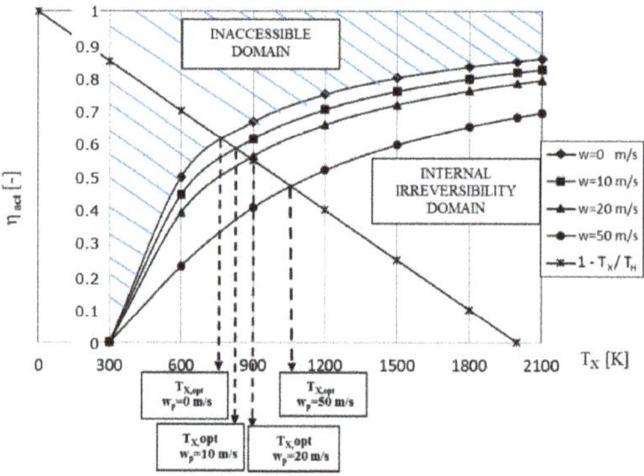

Figure 13. Graphical determination of optimal temperature.

4. Conclusions

Important performance parameters of an irreversible Carnot cycle engine based on optimization models developed in Thermodynamics with Finite Speed and by using the Direct Method have been presented. This analysis predicts lower values of Carnot cycle efficiency than is predicted by the Thermodynamics in Finite Time (FTT), as originated by Chambadal and Curzon–Ahlborn. The piston speed for maximum power and for maximum efficiency has been found for two sets of engine parameters, and it has been shown that entropy generation per time clearly differs from entropy generation per cycle. Moreover, a minimum occurs for the entropy generation per cycle at optimum piston speed corresponding to maximum power.

This study produces a more realistic model for the design of Carnot cycle engines since it includes many of the various internal and external irreversible processes that occur in the actual operation of these engines and correlates them with the finite speed of the piston.

The present analysis has shown that the first unification attempt of TFS and FTT optimization involves analytical correction of the Curzon–Ahlborn efficiency, which is well known as a nice radical, by a term accounting for internal irreversibilities of the Carnot cycle engine. They were evaluated based on the Fundamental Equation of TFS, the First Law for Processes with Finite Speed, where the main irreversibility causes are accounted for, namely, finite speed of the piston, friction, and throttling. This correction appears not only in the Carnot cycle efficiency but also in the optimum temperature of the gas at the hot end of the engine for maximum power, and in the non-dimensional power output of the engine. Thus, the engine performances were derived analytically for a Carnot engine with external and internal irreversibilities generated by finite speed w.

A step further in this first unification approach did a comparison between TFS and FTT optimization results for a Carnot cycle emphasizing that TFS analysis can account for both kind of irrevesibilities, and it can also provide improvement of FTT results.

Thermodynamic analysis based on the Direct Method and Finite Speed of the processes is shown to be especially effective for engineering optimizations since the efficiency and power can each be optimized based on gas temperatures and process speed. The fact that it is already used by other researchers [54–58,84–87] proves its capability to become a useful tool in thermal machine analysis and optimization.

We do hope that this work marks an important step toward the development of a more powerful Engineering Irreversible Thermodynamics, which could be a synthesis unifying the three important branches, namely Thermodynamics with Finite Speed, Thermodynamics with Finite Dimensions, and Thermodynamics in Finite Time.

Author Contributions: M.C. contributed to the development of the model, synthesis, and preparation of the manuscript; S.P. contributed substantially to the development of the model and interpretation of the results; M.F. contributed to the analysis and interpretation of the results; C.D. contributed to the development of the model and analysis of the results; B.B. contributed to the development of the model and results illustration. All authors have read and agreed to the published version of the manuscript.

Funding: This research received no external funding.

Institutional Review Board Statement: Not applicable.

Informed Consent Statement: Not applicable.

Data Availability Statement: The data presented in this study are available on request from the corresponding author.

Conflicts of Interest: The authors declare no conflict of interest.

References

1. Salamon, P. Some Issues in Finite Time Thermodynamics. In *Thermodynamic Optimization and Complex. Energy Systems*; Bejan, A., Mamut, E., Eds.; NATO Science Series, 3. High Technology; Kluwer Academic Publishers: Amsterdam, The Netherlands, 1999; Volume 69, pp. 421–424.
2. Bejan, A. *Advanced Engineering Thermodynamics*; Wiley: New York, NY, USA, 1988.
3. Curzon, F.L.; Ahlborn, B. Efficiency of a carnot engine at maximum power output. *Am. J. Phys.* **1975**, *43*, 22–24. [CrossRef]
4. Chambadal, P. *Les Centrales Nucléaires*; A. Colin: Paris, France, 1957.
5. Novikov, I.I. The efficiency of atomic power stations (a review). In *Journal of Nuclear Energy II*; Pergamon Press Ltd.: London, UK, 1958; Volume 7, pp. 125–128.
6. Gyftopulous, E.P. Fundamentals of analysis of processes. *Energy Convers. Manag.* **1997**, *38*, 1525–1533. [CrossRef]
7. Moran, M.J. On second law analysis and the failed promises of finite time thermodynamics. *Energy* **1998**, *23*, 517–519. [CrossRef]
8. Sekulic, D.P. A fallacious argument in the finite time thermodynamics concept of endoreversibilit. *J. Appl. Phys.* **1998**, *83*, 4561–4565. [CrossRef]
9. Gyftopoulos, E.P. Infinite time (reversible) versus finite time (irreversible) thermodynamics: A misconceived distinction. *Energy* **1999**, *24*, 1035–1039. [CrossRef]
10. Ishida, M. The role and limitations of endoreversible thermodynamics. *Energy* **1999**, *24*, 1009–1014. [CrossRef]
11. Gyftopoulos, E.P. On the Curzon-Ahlborn efficiency and its lack of connection to power producing processes. *Energy Convers. Manag.* **2002**, *43*, 609–615. [CrossRef]
12. Kodal, A.; Sahin, B.; Yilmaz, T.A. Comparative Performance analysis of irreversible Carnot engine under maximum power density and maximum power conditions. *Energy Convers. Manag.* **2000**, *41*, 235–248. [CrossRef]
13. Zhou, S.; Chen, L.; F Sun, F.; Wu, C. Optimal Performance of A Generalized Irreversible Carnot-Engine. *Appl. Energy* **2005**, *81*, 376–387. [CrossRef]
14. Ladino-Luna, D. On optimization of a non-endoreversible curzon-ahlborn cycle. *Entropy* **2007**, *9*, 186–197. [CrossRef]
15. Nie, W.; VHe, J.; Deng, X. Local stability analysis of an irreversible Carnot heat engine. *Int. J. Therm. Sci.* **2008**, *47*, 633–640. [CrossRef]
16. Ibrahim, O.M.; Klein, S.A.; Mitchell, J.W. Economic Evaluation of the Maximum Power Efficiency Concept, ASME Winter Annual Meeting, Atlanta, Georgia, USA, 1–6 December 1991. *J. Eng. Gas. Turbine Power* **1991**, *113*, 514. [CrossRef]
17. Wu, C.; Kiang, R. Finite-time thermodynamics analysis of a Carnot engine with irreversibility. *Energy* **1992**, *17*, 1173–1178. [CrossRef]

18. Feidt, M. Optimal Thermodynamics—New Upperbounds. *Entropy* **2009**, *11*, 529–547. [CrossRef]
19. Salamon, P. Physics versus engineering of finite-time thermodynamic models and optimizations. In *Thermodynamic Optimization of Complex Energy Systems, NATO Advanced Study Institute, Neptun, Romania, 13–24 July 1998*; Bejan, A., Mamut, E., Eds.; Kluwer Academic Publishers: Amsterdam, The Netherlands, 1999; pp. 421–424.
20. Salamon, P. A contrast between the physical and the engineering approaches to finite-time thermodynamic models and optimizations. In *Recent Advances in Finite Time Thermodynamics*; Wu, C., Chen, L., Chen, J., Eds.; Nova Science Publishers: New York, NY, USA, 1999; pp. 541–552.
21. Andresen, B. Comment on "A fallacious argument in the finite time thermodynamics concept of endoreversibility". *J. Appl. Phys.* **2001**, *90*, 6557–6559. [CrossRef]
22. Salamon, P.; Nulton, J.D.; Siragusa, G.; Andresen, T.R.; Limon, A. Principles of control thermodynamics. *Energy* **2001**, *26*, 307–319. [CrossRef]
23. Chen, J.C.; Yan, Z.J.; Lin, G.X.; Andresen, B. On the Curzon-Ahlborn efficiency and its connection with the efficiencies of real heat engines. *Energy Convers. Manag.* **2001**, *42*, 173–181. [CrossRef]
24. Tsirlin, A.; Sukin, I. Averaged optimization and finite-time thermodynamics. *Entropy* **2020**, *22*, 912. [CrossRef] [PubMed]
25. Gonzales-Ayala, J.G.; Roco, J.M.M.; Medina, A.; Hernandez, A.C. optimization, stability, and entropy in endoreversible heat engines. *Entropy* **2020**, *22*, 1323. [CrossRef] [PubMed]
26. Muschik, W.; Hoffmann, K.H. Modeling, simulation, and reconstruction of 2-reservoir heat-to-power processes in finite-time thermodynamics. *Entropy* **2020**, *22*, 997. [CrossRef] [PubMed]
27. Insinga, A.R. The quantum friction and optimal finite-time performance of the quantum otto cycle. *Entropy* **2020**, *22*, 1060. [CrossRef]
28. Masser, R.; Khodja, A.; Scheunert, M.; Schwalbe, K.; Fischer, A.; Paul, R.; Hoffmann, K.H. Optimized piston motion for an alpha-type stirling engine. *Entropy* **2020**, *22*, 700. [CrossRef]
29. Abiuso, P.; Miller, H.J.D.; Perarnau-Llobet, M.; Scandi, M. Geometric optimization of quantum thermodynamic processes. *Entropy* **2020**, *22*, 1076. [CrossRef]
30. Feng, H.J.; Qin, W.X.; Chen, L.G.; Cai, C.G.; Ge, Y.L.; Xia, S.J. Power output, thermal efficiency and exergy-based ecological performance optimizations of an irreversible KCS-34 coupled to variable temperature heat reservoirs. *Energy Convers. Manag.* **2020**, *205*, 112424. [CrossRef]
31. Tang, C.Q.; Chen, L.G.; Feng, H.J.; Ge, Y.L. Four-objective optimization for an irreversible closed modified simple Brayton cycle. *Entropy* **2021**, *23*, 282. [CrossRef]
32. Chen, L.; Feng, H.; Ge, Y. Power and efficiency optimization for open combined regenerative brayton and inverse brayton cycles with regeneration before the inverse cycle. *Entropy* **2020**, *22*, 677. [CrossRef]
33. Andresen, B. *Finite-Time Thermodynamics*; Physics Laboratory II; University of Copenhagen: Copenhagen, Denmark, 1983.
34. Şahin, B.; Kodal, A.; Yavuz, H. Maximum power density foe an endoreversible Carnot heat engine. *Energy* **1996**, *21*, 1219–1225. [CrossRef]
35. Agnew, B.; Anderson, A.; Frost, T.H. Optimization of a steady-flow Carnot cycle with external irreversibilities for maximum specific output. *Appl. Therm. Eng.* **1997**, *17*, 3–15. [CrossRef]
36. Bojic, M. Cogeneration of power and heat by using endoreversible Carnot engine. *Energy Convers. Manag.* **1997**, *38*, 1877–1880. [CrossRef]
37. Qin, X.; Chen, L.; Sun, F.; Wu, C. Frequency-dependent performance of an endoreversible Carnot engine with a linear phenomenological heat-transfer. *Appl. Energy* **2005**, *81*, 365–375.
38. Ding, Z.; Chen, L.; Sun, F. Finite time exergoeconomic performance for six endoreversible heat engine cycles: Unified description. *Appl. Math. Model.* **2011**, *35*, 728–736. [CrossRef]
39. Tierney, M. Minimum exergy destruction from endoreversible and finite-time thermodynamics machines and their concomitant indirect energy. *Energy* **2020**, *197*, 117184. [CrossRef]
40. Feidt, M. *Thermodynamique et Optimisation Energetique des Systèmes et Procédés*, 2nd ed.; Technique et Documentation; Lavoisier: Paris, France, 1996.
41. Chen, J. The maximum power output and maximum efficiency of an irreversible carnot heat engine. *J. Phys. D Appl. Phys.* **1994**, *27*, 1144–1149. [CrossRef]
42. Chen, L.; Sun, F.; Wu, C. A generalized model of real heat engines and its performance. *J. Inst. Energy* **1996**, *69*, 214–222.
43. Feidt, M.; Costea, M.; Petrescu, S.; Stanciu, C. Nonlinear thermodynamic analysis and optimization of a carnot engine cycle. *Entropy* **2016**, *18*, 243. [CrossRef]
44. Feidt, M. Thermodynamics of energy systems and processes: A review and perspectives. *J. Appl. Fluid Mech.* **2012**, *5*, 85–98.
45. Feidt, M. *Thermodynamique Optimale en Dimensions Physiques Finies*; Lavoisier: Paris, France, 2013.
46. Petrescu, S.; Stanescu, G.; Costea, M. The study for optimization of the Carnot cycle which develops with finite speed. In Proceedings of the International Conference on Energy Systems and Ecology, ENSEC'93, Cracow, Poland, 5–8 July 1993; Volume 1, pp. 269–277.
47. Petrescu, S.; Harman, C.; Bejan, A. The Carnot cycle with external and internal irreversibilities. In Proceedings of the Florence World Energy Research Symposium, Energy for the 21st Century: Conversion, Utilization and Environmental Quality, Firenze, Italy, 6–8 July 1994.

48. Petrescu, S.; Feidt, M.; Harman, C.; Costea, M. Optimization of the irreversible Carnot cycle engine for maximum efficiency and maximum power through use of finite speed thermodynamic analysis. In Proceedings of the ECOS'2002 Conference, Berlin, Germany, 3–5 July 2002; Tsatsaronis, G., Moran, M., Cziesla, F., Bruckner, T., Eds.; Volume 2, pp. 1361–1368.

49. Petrescu, S.; Harman, C.; Costea, M.; Feidt, M. Thermodynamics with Finite Speed versus Thermodynamics in Finite Time in the Optimization of Carnot Cycle. In Proceedings of the 6-th ASME-JSME Thermal Engineering Joint Conference, Hawaii, HI, USA, 16–20 March 2003.

50. Petre, C. Utilizarea Termodinamicii cu Viteza Finita in Studiul si Optimizarea Ciclului Carnot si a Masinilor Stirling (The Use of Thermodynamics with Finite Speed to the Study and Optimization of Carnot Cycle and Stirling Machines). Ph.D. Thesis, University Politehnica of Bucharest, Bucharest, Romania, University H. Poincaré of Nancy, Nancy, France, 2007.

51. Feidt, M.; Costea, M.; Petre, C.; Petrescu, S. Optimization of direct Carnot cycle. *Appl. Therm. Eng.* **2007**, *27*, 829–839. [CrossRef]

52. Petrescu, S.; Harman, C.; Bejan, A.; Costea, M.; Dobre, C. Carnot cycle with external and internal irreversibilities analyzed in thermodynamics with finite speed with the direct method. *Termotehnica* **2011**, *15*, 7–17.

53. Liu, X.W.; Chen, L.G.; Wu, F.; Sun, F.R. Optimal performance of a spin quantum Carnot heat engine with multi-irreversibilities. *J. Energy Inst.* **2014**, *87*, 69–80. [CrossRef]

54. Feng, H.J.; Chen, L.G.; Sun, F.R. Optimal Ratios of the Piston Speeds for a Finite Speed Endoreversible Carnot Heat Engine Cycle. *Rev. Mex. Fis.* **2010**, *56*, 135–140.

55. Feng, H.J.; Chen, L.G.; Sun, F.R. Optimal ratio of the piston for a finite speed irreversible carnot heat engine cycle. *Int. J. Sustain. Energy* **2010**, *30*, 321–335. [CrossRef]

56. Feng, H.J.; Chen, L.G.; Sun, F.R. Effects of unequal finite speed on the optimal performance of endoreversible Carnot refrigeration and heat pump cycles. *Int. J. Sustain. Energy* **2011**, *30*, 289–301. [CrossRef]

57. Yang, B.; Chen, L.G.; Sun, F.R. Performance analysis and optimization for an endoreversible Carnot heat pump cycle with finite speed of piston. *Int. J. Energy Environ.* **2011**, *2*, 1133–1140.

58. Chen, L.G.; Feng, H.J.; Sun, F.R. Optimal piston speed ratios for irreversible carnot refrigerator and heat pump using finite time thermodynamics, finite speed thermodynamics and the direct method. *J. Energy Inst.* **2011**, *84*, 105–112. [CrossRef]

59. Petrescu, S.; Harman, C.; Costea, M.; Florea, T.; Petre, C. *Advanced Energy Conversion*; Bucknell University: Lewisburg, PA, USA, 2006.

60. Petrescu, S.; Costea, M.; Petrescu, V.; Malancioiu, O.; Boriaru, N.; Stanciu, C.; Banches, E.; Dobre, C.; Maris, V.; Leontiev, C. *Development of Thermodynamics with Finite Speed and Direct Method*; AGIR: Bucharest, Romania, 2011.

61. Petrescu, S.; Costea, M.; Feidt, M.; Ganea, I.; Boriaru, N. *Advanced Thermodynamics of Irreversible Processes with Finite Speed and Finite Dimensions. A Historical and Epistemological Approach, with Extension to Biological and Social Systems*; AGIR: Bucharest, Romania, 2015.

62. Costea, M.; Petrescu, S.; Harman, C. The effect of irreversibilities on solar stirling engine cycle performance. *Energy Convers. Manag.* **1999**, *40*, 1723–1731. [CrossRef]

63. Petrescu, S.; Costea, M.; Harman, C.; Florea, T. Application of the direct method to irreversible stirling cycles with finite speed. *Int. J. Energy Res.* **2002**, *26*, 589–609. [CrossRef]

64. Florea, T.; Petrescu, S.; Florea, E. *Schemes for Computation and Optimization of the Irreversible Processes in Stirling Machines*; Leda & Muntenia: Constanta, Romania, 2000.

65. Petrescu, S.; Cristea, A.F.; Boriaru, N.; Costea, M.; Petre, C. Optimization of the Irreversible Otto Cycle using Finite Speed Thermodynamics and the Direct Method. In Proceedings of the 10th WSEAS Int. Conf. on Mathematical and Computational Methods in Science and Engineering (MACMESE'08), Computers and Simulation in Modern Science, Bucharest, Romania, 7–9 November 2008; Mastorakis, N., Ed.; World Scientific and Engineering Academy and Society (WSEAS): Stevens Point, WI, USA; Volume 2, pp. 51–56.

66. Cullen, B.; McGovern, J.; Petrescu, S.; Feidt, M. Preliminary modelling results for otto-stirling hybrid. cycle. In Proceedings of the ECOS 2009, Foz de Iguasu, Parana, Brazil, 31 August–3 September 2009; pp. 2091–2100.

67. McGovern, J.; Cullen, B.; Feidt, M.; Petrescu, S. Validation of a simulation model for a combined otto and stirling cycle power plant. In Proceedings of the ASME 2010, 4th International Conference on Energy Sustainability, ES2010, Phoenix, AZ, USA, 17–22 May 2010.

68. Petrescu, S.; Boriaru, N.; Costea, M.; Petre, C.; Stefan, A.; Irimia, C. Optimization of the irreversible diesel cycle using finite speed thermodynamics and the direct method. *Bull. Transilv. Univ. Braşov* **2009**, *2*, 87–94.

69. Petrescu, S.; Petrescu, V.; Stanescu, G.; Costea, M. A Comparison between Optimization of Thermal Machines and Fuel Cells based on New Expression of the First Law of Thermodynamics for Processes with Finite Speed. In Proceedings of the 1st Conference on Energy ITEC' 93, Marrakesh, Morocco, 6–10 June 1993; pp. 654–657.

70. Petrescu, S. Contributions to the Study of Interactions and Processes of Non-Equilibrium in Thermal Machine. Ph.D. Thesis, Polytechnic Institute of Bucharest, Bucharest, Romania, 1969.

71. Stoicescu, L.; Petrescu, S. The First Law of Thermodynamics for Processes with Finite Constant Speed in Closed Systems. *Polytech. Inst. Buchar. Bull.* **1964**, *26*, 87–108.

72. Stoicescu, L.; Petrescu, S. Thermodynamic processes developing with constant finite speed. *Polytech. Inst. Buchar. Bull.* **1964**, *26*, 79–119.

73. Stoicescu, L.; Petrescu, S. Thermodynamic processes developing with variable finite speed. *Polytech. Inst. Buchar. Bull.* **1965**, *27*, 65–96.
74. Stoicescu, L.; Petrescu, S. Thermodynamic cycles with finite speed. *Polytech. Inst. Buchar. Bull.* **1965**, *27*, 82–95.
75. Petrescu, S. *Lectures on New Sources of Energy*; Helsinki University of Technology: Otaniemi, Finland, 1991.
76. Petrescu, S.; Iordache, R.; Stanescu, G.; Dobrovicescu, A. The First Law of Thermodynamics for Closed Systems, Considering the Irreversibilities Generated by Friction Piston-Cylinder, the Throttling of the Working Medium and the Finite Speed of Mechanical Interaction. In Proceedings of the ECOS'92, Zaragoza, Spain, 15–19 June 1992.
77. Petrescu, S.; Stanescu, G. The direct method for studying the irreversible processes undergoing with finite speed in closed systems. *Termotehnica* **1993**, *1*, 69–84.
78. Petrescu, S.; Harman, C. The Connection between the First and Second Law of Thermodynamics for Processes with Finite Speed. A Direct Method for Approaching and Optimization of Irreversible Processes. *J. Heat Transf. Soc. Jpn.* **1994**, *33*, 60–67.
79. Stanescu, G. The Study of the Mechanism of Irreversibility Generation in Order to Improve the Performances of Thermal Machines and Devices. Ph.D. Thesis, U.P.B., Bucharest, Romania, 1993.
80. Petrescu, S.; Petre, C.; Costea, M.; Malancioiu, O.; Boriaru, N.; Dobrovicescu, A.; Feidt, M.; Harman, C. A methodology of computation, design and optimization of solar stirling power plant using hydrogen/oxygen fuel cells. *Energy* **2010**, *35*, 729–739. [CrossRef]
81. Heywood, J.B. *Internal Combustion Engine Fundamentals*; McGraw-Hill: New York, NY, USA, 1988.
82. Hagen, K.D. *Heat Transfer with Applications*; Prentice Hall Inc.: Upper Saddle River, NJ, USA, 1999.
83. Costea, M. Improvement of Heat Exchangers Performance in View of the Thermodynamic Optimization of Stirling Machine; Unsteady-State Heat Transfer in Porous Media. Ph.D. Thesis, University Politehnica of Bucharest, Bucharest, Romania; University H. Poincaré of Nancy, Nancy, France, 1997.
84. Hosseinzade, H.; Sayyaadi, H.; Babaelahi, M. A new closed-form analytical thermal model for simulating Stirling engines based on polytropic—Finite speed thermodynamics. *Energy Convers. Manag.* **2015**, *90*, 395–408. [CrossRef]
85. Hosseinzade, H.; Sayyaadi, H. CAFS: The combined adiabatic–finite speed thermal model for simulation and optimization of stirling engines. *Energy Convers. Manag.* **2015**, *91*, 32–53. [CrossRef]
86. Ahmadi, M.H.; Ahmadi, M.A.; Pourfayaz, F.; Bidi, M.; Hosseinzade, H.; Feidt, M. Optimization of powered stirling heat engine with finite speed thermodynamics. *Energy Convers. Manag.* **2016**, *108*, 96–105. [CrossRef]
87. Wu, H.; Ge, Y.L.; Chen, L.G.; Feng, H.J. Power, efficiency, ecological function and ecological coefficient of performance optimizations of irreversible Diesel cycle based on finite piston speed. *Energy* **2021**, *216*, 119235. [CrossRef]

Article

Power and Thermal Efficiency Optimization of an Irreversible Steady-Flow Lenoir Cycle

Ruibo Wang [1,2], **Yanlin Ge** [1,2], **Lingen Chen** [1,2,*], **Huijun Feng** [1,2] **and Zhixiang Wu** [1,2]

[1] Institute of Thermal Science and Power Engineering, Wuhan Institute of Technology, Wuhan 430205, China; ruibowq@126.com (R.W.); geyali9@hotmail.com (Y.G.); huijunfeng@139.com (H.F.); zhixiangwuhg@outlook.com (Z.W.)

[2] School of Mechanical & Electrical Engineering, Wuhan Institute of Technology, Wuhan 430205, China

* Correspondence: lgchenna@yahoo.com

Abstract: Using finite time thermodynamic theory, an irreversible steady-flow Lenoir cycle model is established, and expressions of power output and thermal efficiency for the model are derived. Through numerical calculations, with the different fixed total heat conductances (U_T) of two heat exchangers, the maximum powers (P_{max}), the maximum thermal efficiencies (η_{max}), and the corresponding optimal heat conductance distribution ratios ($u_{L_P(opt)}$) and ($u_{L_\eta(opt)}$) are obtained. The effects of the internal irreversibility are analyzed. The results show that, when the heat conductances of the hot- and cold-side heat exchangers are constants, the corresponding power output and thermal efficiency are constant values. When the heat source temperature ratio (τ) and the effectivenesses of the heat exchangers increase, the corresponding power output and thermal efficiency increase. When the heat conductance distributions are the optimal values, the characteristic relationships of $P - u_L$ and $\eta - u_L$ are parabolic-like ones. When U_T is given, with the increase in τ, the P_{max}, η_{max}, $u_{L_P(opt)}$, and $u_{L_\eta(opt)}$ increase. When τ is given, with the increase in U_T, P_{max} and η_{max} increase, while $u_{L_P(opt)}$ and $u_{L_\eta(opt)}$ decrease.

Keywords: finite time thermodynamics; irreversible Lenoir cycle; cycle power; thermal efficiency; heat conductance distribution; performance optimization

Citation: Wang, R.; Ge, Y.; Chen, L.; Feng, H.; Wu, Z. Power and Thermal Efficiency Optimization of an Irreversible Steady-Flow Lenoir Cycle. *Entropy* **2021**, *23*, 425. https://doi.org/10.3390/e23040425

Academic Editor: Michel Feidt

Received: 15 March 2021
Accepted: 31 March 2021
Published: 2 April 2021

Publisher's Note: MDPI stays neutral with regard to jurisdictional claims in published maps and institutional affiliations.

1. Introduction

Finite time thermodynamic (FTT) theory [1–4] has been applied to the performance analysis and optimization of heat engine (HEG) cycles, and fruitful results have been achieved for both reciprocating and steady-flow cycle models. For the steady-flow models, FTT was also termed as finite physical dimensions thermodynamics by Feidt [5–10]. The famous thermal efficiency formula $\eta = 1 - \sqrt{T_L/T_H}$, where T_H and T_L are the temperatures of the heat source and heat sink of a HEG, was derived by Moutier [11] in 1872, Cotterill [12] in 1890, and Novikov [13] and Chambadel [14] in 1957 for steady-flow power plants, while the systematical analysis combining thermodynamics with heat transfer for Carnot cycle was performed by Curzon and Ahlborn [15] in 1975 for reciprocating model, and FTT development was promoted by Berry's group [4].

A large number of works have been performed for reciprocating (finite time) models [16–25] by applying FTT. While finite size is the major feature for steady-flow devices, such as closed gas rubine (Brayton cycle) power plants and steam (Rankine cycle) and organic Rankine cycle power plants, many scholars have performed FTT studies for various steady-flow cycles with the power output (POW), thermal efficiency (TEF), exergy efficiency, profit rate, and ecological function as the optimization goals, under the conditions of different losses and heat transfer laws [26–51].

Lenoir [52] first proposed the Lenoir cycle (LC) model in 1860. The simple LC consists of only three processes of constant-volume endothermic, adiabatic expansion, and constant-pressure exothermic; the LC is also called the triangular cycle. According to the cycle

59

form, LC can be divided into steady-flow and reciprocating. Georgiou [53] first used classical thermodynamics to study the performances of simple, regenerated, and modified regenerated steady-flow Lenoir cycles (SFLCs).

Following on from [53], Shen et al. [54] applied FTT theory to optimize the POW and TEF characteristics of the endoreversible SFLC with only the loss of heat resistance, and they studied the influences of heat source temperature ratio and total heat conductance (HC) on cycle performance. Ahmadi et al. [55] used a genetic algorithm to carry out multiobjective optimization for endoreversible SFLC, and they obtained the optimal values of ecological performance coefficient and thermal economy under different temperature ratios.

In this paper, an irreversible SFLC model will be established on the basis of [54], while the cycle performance will be analyzed and optimized with the POW and TEF as objective functions, the optimal HC distributions of hot- and cold-side heat exchangers (HACHEX) of the cycle will be studied under different fixed total HCs, and the characteristic relationships between POW and TEF versus HC distribution are obtained. The effect of the internal irreversibility will be analyzed.

2. Cycle Model

Figures 1 and 2 show the $T - s$ and $p - v$ diagrams of the irreversible SFLC. As can be seen, $1 \rightarrow 2$ is the constant-volume endothermic process, $2 \rightarrow 3$ is the irreversible adiabatic expansion process ($2 \rightarrow 3S$ is the corresponding isentropic process), and $3 \rightarrow 1$ is the constant-pressure exothermic process. Assuming the cycle WF is an ideal gas, the entire cycle needs to be completed between the heat source (T_H) and heat sink (T_L).

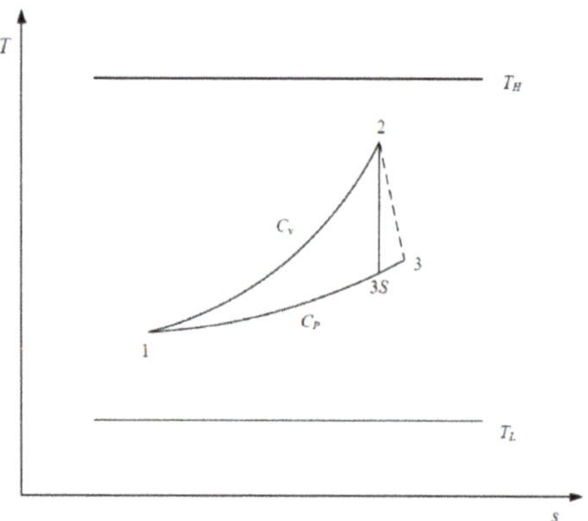

Figure 1. $T - s$ diagram for the irreversible steady-flow Lenoir cycle (SFLC).

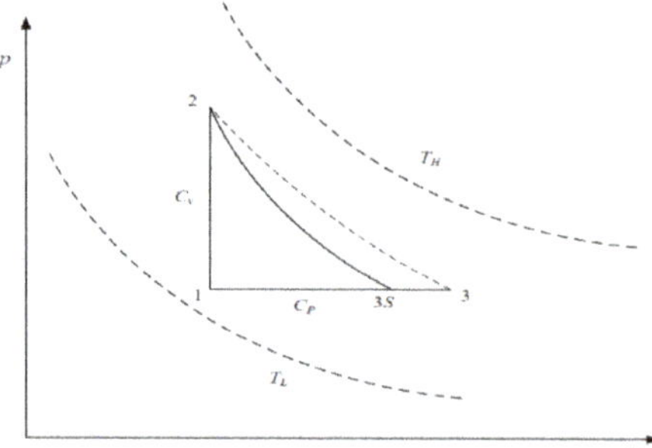

Figure 2. $p - v$ diagram for the irreversible SFLC.

In the actual work of the HEG, there are irreversible losses during compression and expansion processes; thus, the irreversible expansion efficiency η_E is defined to describe the irreversible loss during the expansion process.

$$\eta_E = \frac{T_2 - T_3}{T_2 - T_{3S}}, \tag{1}$$

where T_i $(i = 2, 3, 3S)$ is the corresponding state point temperature.

Assuming that the heat transfer between the WF and heat reservoir obeys the law of Newton heat transfer, according to the theory of the heat exchanger (HEX) and the ideal gas properties, the cycle heat absorbing and heat releasing rates are, respectively,

$$\dot{Q}_{1\to2} = \dot{m}C_v E_H (T_H - T_1) = \dot{m}C_v (T_2 - T_1), \tag{2}$$

$$\dot{Q}_{3\to1} = \dot{m}C_P E_L (T_3 - T_L) = \dot{m}C_P (T_3 - T_1), \tag{3}$$

where $\dot{m}$ is the mass flow rate of the WF, $C_v(C_P)$ is the constant-volume (constant-pressure) SH ($C_P = kC_v$, k is the cycle SH ratio), and $E_H(E_L)$ is the effectiveness of hot-side (cold-side) HEX.

The relationships among the effectivenesses with the corresponding heat transfer unit numbers (N_H, N_L) and HCs (U_H, U_L) are as follows:

$$N_H = U_H / (\dot{m}C_v), \tag{4}$$

$$N_L = U_L / (\dot{m}kC_v), \tag{5}$$

$$E_H = 1 - \exp(-N_H), \tag{6}$$

$$E_L = 1 - \exp(-N_L). \tag{7}$$

3. Analysis and Discussion

3.1. Power and Thermal Efficiency Expressions

According to the second law of thermodynamics, after a cycle process, the total entropy change of the WF is equal to zero; thus, one finds

$$C_v \ln(T_2 / T_1) - C_P \ln(T_{3S} / T_1) = 0. \tag{8}$$

From Equation (8), one obtains

$$\frac{T_2}{T_1} = \left(\frac{T_{3S}}{T_1}\right)^k. \tag{9}$$

From Equations (2) and (3), one has

$$T_2 = E_H (T_H - T_1) + T_1, \tag{10}$$

$$T_3 = (E_L T_L - T_1)/(E_L - 1). \tag{11}$$

Combining Equations (1), (9), and (10) with Equation (11) yields

$$T_1 = \frac{E_H T_H (\eta_E - 1) + (T_1 - E_L T_L)/(1 - E_L)}{\{(1 - E_H)(1 - \eta_E) + \{[E_H T_H + (1 - E_H)T_1]/T_1\}^{\frac{1}{k}} \eta_E\}}. \tag{12}$$

From Equations (2), (3) and (9)–(11), the POW and TEF expressions of the irreversible SFLC can be obtained as

$$P = \dot{Q}_{1\to2} - \dot{Q}_{3\to1} = \dot{m}C_v \left[E_H (T_H - T_1) - \frac{k E_L (T_1 - T_L)}{1 - E_L}\right], \tag{13}$$

$$\eta = P/\dot{Q}_{1\to 2} = 1 - \frac{kE_L(T_1 - T_L)}{E_H(1 - E_L)(T_H - T_1)}. \tag{14}$$

When $\eta_E = 1$, Equation (12) simplifies to

$$T_1 - E_L T_L = (1 - E_L)[E_H T_H + (1 - E_H)T_1]^{\frac{1}{k}} T_1^{1-\frac{1}{k}}. \tag{15}$$

Equation (15) in this paper is consistent with Equation (15) in [54], where T_1 was obtained for the endoreversible SFLC. Combining Equations (13)–(15) and using the numerical solution method, the POW and TEF characteristics of the endoreversible SFLC in [54] can be obtained.

3.2. Case with Given Hot- and Cold-Side HCs

The working cycles of common four-branch HEGs, such as Carnot, Brayton, and Otto engines, can be roughly divided into four processes: compression, endothermic, expansion, and exothermic. Compared with these common four-stroke cycles, the biggest feature of the SFLC is the lack of a gas compression process, presenting a relatively rare three-branch cycle model.

When the hot- and cold-side HCs are constant, it can be seen from Equations (4)–(7) that the effectivenesses of the HACHEX which are directly related to each cycle state point temperature will be fixed values; as a result, the POW and TEF will also be fixed values.

3.3. Case with Variable Hot- and Cold-Side HCs When Total HC Is Given

When the HC changes, the POW and TEF of the cycle will also change; therefore, the HC can be optimized and the optimal POW and TEF can be obtained. Assuming the total HC is a constant,

$$U_L + U_H = U_T. \tag{16}$$

Defining the HC distribution ratio as $u_L = \frac{U_L}{U_T}(0 < u_L < 1)$, from Equations (4)–(7), the effectivenesses of the HACHEX can be represented as

$$E_H = 1 - \exp[-(1 - u_L)U_T/(\dot{m}C_v)], \tag{17}$$

$$E_L = 1 - \exp[-u_L U_T/(\dot{m}kC_v)]. \tag{18}$$

Combining Equations (12)–(14) and (17) with Equation (18) and using a numerical solution method, the characteristic relationships between POW and the hot- and cold-side HC distribution ratio, as well as between TEF and the hot- and cold-side HC distribution ratio, can be obtained.

4. Numerical Examples

It is assumed that the working fluid is air. Therefore, its constant-volume specific heat and specific heat ratio are $C_v = 0.7165 \text{ kJ}/(\text{kg·K})$ and $k = 1.4$. The turbine efficiency of the gas turbine is about $\eta_E = 0.92$ in general. According to the [51–55], $\dot{m} = 1.1165 \text{ kg/s}$ and $T_L = 320 \text{ K}$ were set.

Figure 3 shows the POW and TEF characteristics when the HCs of the HACHEX and temperature ratio are different values. When the HCs and temperature ratio are fixed values, the effectivenesses of the HEX are fixed values, and the corresponding POW and TEF are also fixed values. The POW and TEF characteristics are reflected in the graph as a point. As can be seen, when $\tau(\tau = T_H/T_L)$ and the HCs of the HEXs increase, the corresponding POW and TEF increase. Figure 4 shows the influence of η_E on $P - \eta$ characteristics when the HCs of HACHEX and temperature ratio are given. As can be seen, with the increase in η_E (the decrease of irreversible loss), the corresponding P and η increase.

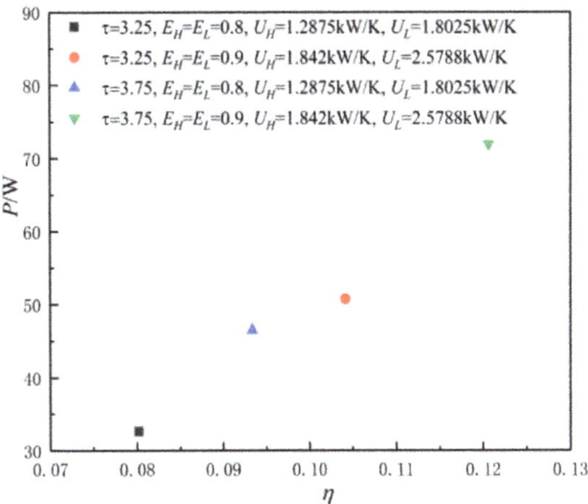

Figure 3. The power output (POW) and thermal efficiency (TEF) characteristics when the HCs of HACHEX are given.

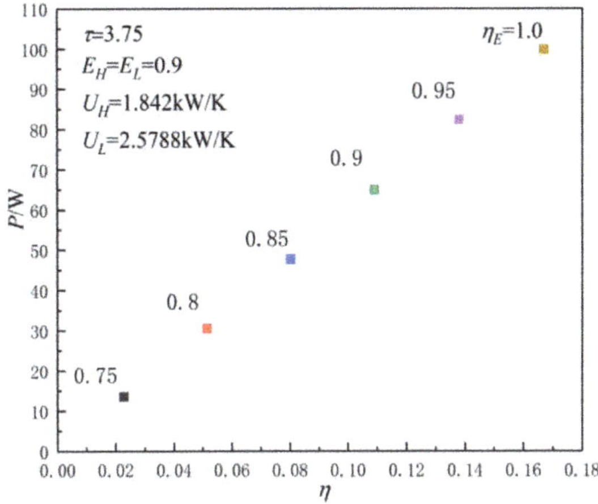

Figure 4. Effect of η_E on $P - \eta$ characteristics when the HCs of HACHEX are given.

Figures 5–8 show the influences of U_T on the $P - u_L$ and $\eta - u_L$ characteristics when $\tau = 3.25$ and $\tau = 3.75$. The relationship curves of $P - u_L$ and $\eta - u_L$ are parabolic-like changes. With the increase in u_L, the corresponding POW and TEF first increase and then decrease, and there are optimal HC distribution values $u_{Lp(opt)}$ and $u_{L\eta(opt)}$, which lead to POW and TEF reaching their maximum values $P_{\max}$ and $\eta_{\max}$.

Figures 5 and 6 show the influence of U_T on $P - u_L$ characteristics when $\tau = 3.25$ and $\tau = 3.75$. As can be seen, with the increase in U_T, $P_{\max}$ increases and $u_{Lp(opt)}$ decreases. When U_T is 2.5, 5, 7.5, and 10 kW/K and $\tau = 3.25$, the corresponding $P_{\max}$ is 23.04, 56.58, 70.25, and 74.39 W, while $u_{Lp(opt)}$ is 0.58, 0.575, 0.574, and 0.573, respectively. When U_T changes from 2.5 to 10 kW/K, the corresponding $P_{\max}$ increases by about 222.9%, while the $u_{Lp(opt)}$ decreases by about 1.21%. When U_T is 2.5, 5, 7.5, and 10 kW/K and $\tau = 3.75$, the corresponding $P_{\max}$ is 33.06, 80.06, 90.24, and 105.06 W, while $u_{Lp(opt)}$ is 0.586, 0.579, 0.5785, and 0.5782, respectively. When U_T changes from 2.5 to 10 kW/K, the corresponding $P_{\max}$ increases by about 217.8%, while the $u_{Lp(opt)}$ decreases by about 1.33%.

Figures 7 and 8 show the influence of U_T on $\eta - u_L$ characteristics when $\tau = 3.25$ and $\tau = 3.75$. As can be seen, with the increase in U_T, $\eta_{\max}$ increases and $u_{L_\eta(opt)}$ decreases. When U_T is 2.5, 5, 7.5, and 10 kW/K and $\tau = 3.25$, the corresponding $\eta_{\max}$ is 0.066, 0.111, 0.126, and 0.1303, while $u_{L_\eta(opt)}$ is 0.629, 0.614, 0.605, and 0.6, respectively. When U_T changes from 2.5 to 10 kW/K, the corresponding $\eta_{\max}$ increases by about 97.4%, while $u_{L_P(opt)}$ decreases by about 4.61%. When U_T is 2.5, 5, 7.5, and 10 kW/K and $\tau = 3.75$, the corresponding $\eta_{\max}$ is 0.0774, 0.129, 0.1458, and 0.1506, while $u_{L_\eta(opt)}$ is 0.644, 0.624, 0.608, and 0.606, respectively. When U_T changes from 2.5 to 10 kW/K, the corresponding $\eta_{\max}$ increases by about 94.6%, while $u_{L_P(opt)}$ decreases by about 5.9%.

From Figures 5–8 and Equations (12)–(14), (17), and (18), one can see that, when τ is given, the POW and TEF are mainly affected by the total HC; with the increase in U_T, the $P_{\max}$ and $\eta_{\max}$ increase. When the total HC is small, the corresponding $P_{\max}$ and $\eta_{\max}$ change more significantly. When the total HC is large, the corresponding $P_{\max}$ and $\eta_{\max}$ change little. When U_T is given, with the increase in τ, the $u_{L_P(opt)}$ and $u_{L_\eta(opt)}$ increase. When τ and U_T are given, the corresponding $u_{L_\eta(opt)} > u_{L_P(opt)}$.

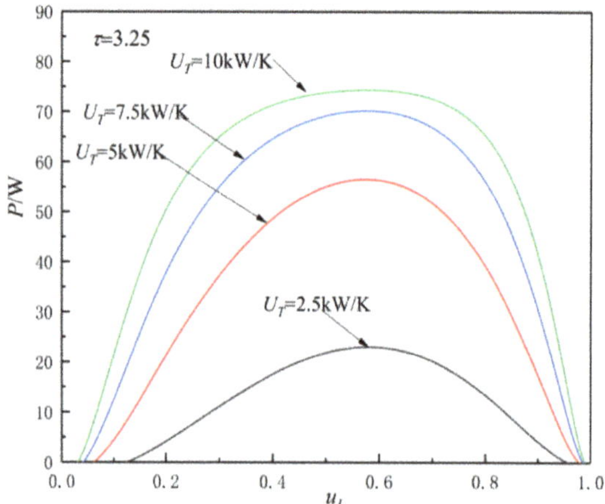

Figure 5. Effect of U_T on $P - u_L$ characteristics when $\tau = 3.25$.

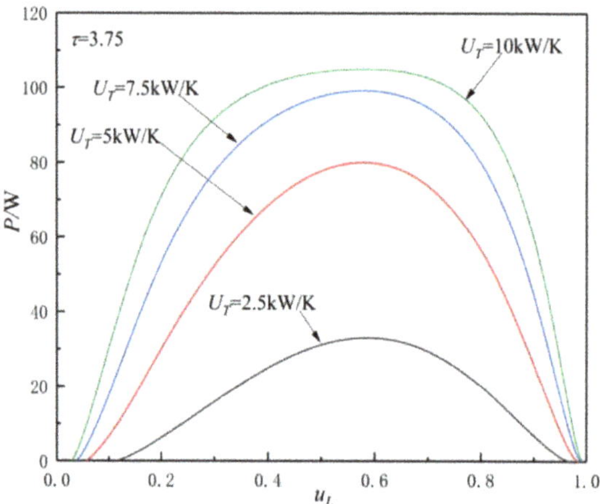

Figure 6. Effect of U_T on $P - u_L$ characteristics when $\tau = 3.75$.

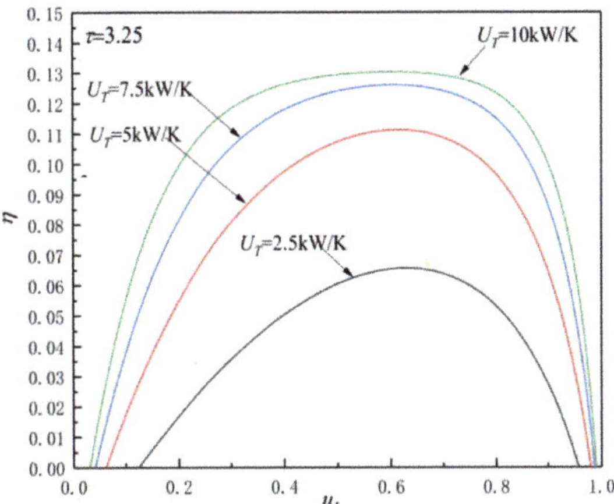

Figure 7. Effect of U_T on $\eta - u_L$ characteristics when $\tau = 3.25$.

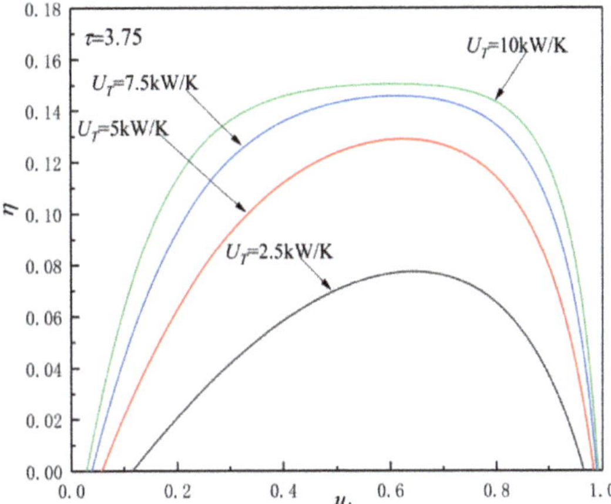

Figure 8. Effect of U_T on $\eta - u_L$ characteristics when $\tau = 3.75$.

Figures 9 and 10 show the influences of η_E on $P - u_L$ and $\eta - u_L$ characteristics when $\tau = 3.75$ and $U_T = 7.5$ kW/K. As can be seen, when $\tau = 3.75$ and $U_T = 7.5$ kW/K, with the increase in η_E (the decrease in irreversible loss), the $P_{\max}$ and $\eta_{\max}$ increase, while the corresponding $u_{L_{P(opt)}}$ and $u_{L_{\eta(opt)}}$ decrease. When η_E is 0.75, 0.8, 0.85, 0.9, 0.95, and 1.0, the corresponding $P_{\max}$ is 30.2431, 50.4808, 70.7674, 91.0982, 111.4719, and 131.8876, $\eta_{\max}$ is 0.0445, 0.0743, 0.1041, 0.1339, 0.1637, and 0.1935, $u_{L_{P(opt)}}$ is 0.601, 0. 593, 0.586, 0.581, 0.576, and 0.572, and $u_{L_{\eta(opt)}}$ is 0.619, 0.617, 0.615, 0.613, 0.611, and 0.609, respectively. When η_E changes from 0.75 to 1.0, the corresponding $P_{\max}$ increases by about 336.1%, $\eta_{\max}$ increases by about 334.8%, $u_{L_{P(opt)}}$, and $u_{L_{\eta(opt)}}$ decreases by about 4.83% and 1.62%, respectively.

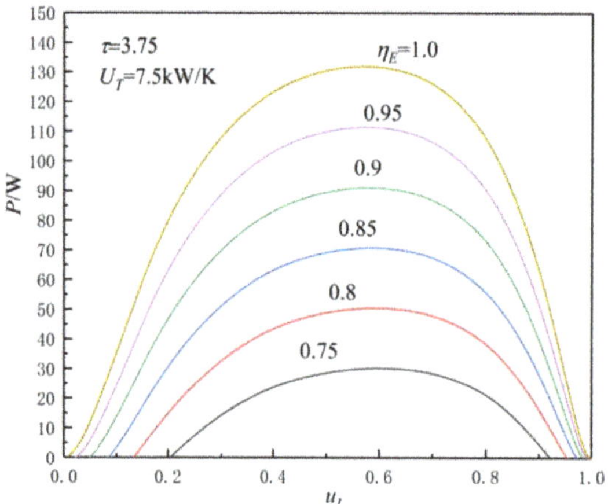

Figure 9. Effect of η_E on $P - u_L$ characteristics.

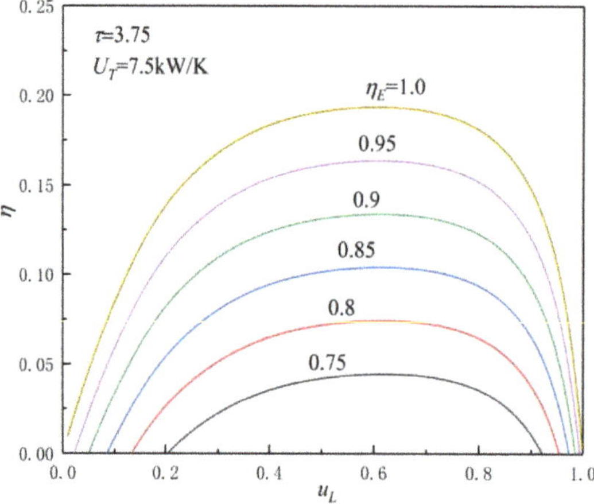

Figure 10. Effect of η_E on $\eta - u_L$ characteristics.

5. Conclusions

In this paper, an irreversible SFLC model is established on the basis of [54], while the POW and TEF characteristics of the irreversible SFLC were studied using FTT theory, and the influences of τ, U_T and η_E on $P_{\max}$, $\eta_{\max}$, $u_{L_P(opt)}$, and $u_{L_\eta(opt)}$ were analyzed. The main conclusions are as follows:

(1) When the HCs are constants, the corresponding POW and TEF are fixed values. When τ and the HCs of the HEXs increase, the corresponding POW and TEF increase. When τ and HCs of the HEXs are constants, with the increase in η_E (the decrease in irreversible loss), the corresponding P and η increase.

(2) When the distribution of HCs can be optimized, the relationships of $P - u_L$ and $\eta - u_L$ are parabolic-like ones.

(3) When U_T is given, with the increase in τ, $P_{\max}$, $\eta_{\max}$, $u_{L_P(opt)}$, and $u_{L_\eta(opt)}$ increase.

(4) When τ is given, with the increase in U_T, $P_{\max}$ and $\eta_{\max}$ increase, while $u_{L_P(opt)}$ and $u_{L_\eta(opt)}$ decrease. When τ and U_T are given, the corresponding $u_{L_\eta(opt)}$ is bigger than $u_{L_P(opt)}$.

(5)   When $\tau = 3.75$ and $U_T = 7.5 \text{kW/K}$, with the increase in η_E, $P_{\max}$ and $\eta_{\max}$ increase, while the corresponding $u_{L_P(opt)}$ and $u_{L_\eta(opt)}$ decrease.

Author Contributions: Conceptualization, R.W. and L.C.; data curation, Y.G.; funding acquisition, L.C.; methodology, R.W., Y.G., L.C. and H.F.; software, R.W., Y.G., H.F. and Z.W.; supervision, L.C.; validation, R.W., H.F. and Z.W.; writing—original draft preparation, R.W. and Y.G.; writing—reviewing and editing, L.C. All authors have read and agreed to the published version of the manuscript.

Funding: This paper is supported by the National Natural Science Foundation of China (Project No. 51779262).

Acknowledgments: The authors wish to thank the reviewers for their careful, unbiased, and constructive suggestions, which led to this revised manuscript.

Conflicts of Interest: The authors declare no conflict of interest.

Nomenclature

C_P	Specific heat at constant pressure (kJ/(kg $\cdot$ K))
C_v	Specific heat at constant volume (kJ/(kg $\cdot$ K))
E	Effectiveness of heat exchanger
k	Specific heat ratio (-)
$\dot{m}$	Mass flow rate of the working fluid (kg/s)
N	Number of heat transfer units
P	Cycle power (W)
$\dot{Q}$	Quantity of heat transfer rate (W)
T	Temperature (K)
U	Heat conductance (kW/K)
U_T	Total heat conductance (kW/K)
u	Heat conductance distribution
Greek symbols	
τ	Temperature ratio
η	Cycle thermal efficiency
Subscripts	
H	Hot-side
L	Cold-side
max	Maximum value
opt	Optimal
P	Maximum power point
η	Maximum thermal efficiency point
$1-3, 3S$	Cycle state points

Abbreviations

FTT	Finite time thermodynamic
HACHEX	Hot- and cold-side heat exchangers
HC	Heat conductance
HEG	Heat engine
HEX	Heat exchanger
LC	Lenoir cycle
POW	Power output
SFLC	Steady flow Lenoir cycle
SH	Specific heat
TEF	Thermal efficiency
WF	Working fluid

References

1.   Andresen, B.; Berry, R.S.; Ondrechen, M.J. Thermodynamics for processes in finite time. *Acc. Chem. Res.* **1984**, *17*, 266–271. [CrossRef]

2. Andresen, B. Current trends in finite-time thermodynamics. *Angew. Chem. Int. Ed.* **2011**, *50*, 2690–2704. [CrossRef]
3. Feidt, M. The history and perspectives of efficiency at maximum power of the Carnot engine. *Entropy* **2017**, *19*, 369. [CrossRef]
4. Berry, R.S.; Salamon, P.; Andresen, B. How it all began. *Entropy* **2020**, *22*, 908. [CrossRef] [PubMed]
5. Feidt, M. *Finite Physical Dimensions Optimal Thermodynamics 1. Fundamental*; ISTE Press and Elsevier: London, UK, 2017.
6. Feidt, M. *Finite Physical Dimensions Optimal Thermodynamics 2. Complex. Systems*; ISTE Press and Elsevier: London, UK, 2018.
7. Blaise, M.; Feidt, M.; Maillet, D. Influence of the working fluid properties on optimized power of an irreversible finite dimensions Carnot engine. *Energy Convers. Manag.* **2018**, *163*, 444–456. [CrossRef]
8. Feidt, M.; Costea, M. From finite time to finite physical dimensions thermodynamics: The Carnot engine and Onsager's relations revisited. *J. Non-Equilib. Thermodyn.* **2018**, *43*, 151–162. [CrossRef]
9. Dumitrascu, G.; Feidt, M.; Popescu, A.; Grigorean, S. Endoreversible trigeneration cycle design based on finite physical dimensions thermodynamics. *Energies* **2019**, *12*, 3165.
10. Feidt, M.; Costea, M.; Feidt, R.; Danel, Q.; Périlhon, C. New criteria to characterize the waste heat recovery. *Energies* **2020**, *13*, 789. [CrossRef]
11. Moutier, J. *Éléments de Thermodynamique*; Gautier-Villars: Paris, France, 1872.
12. Cotterill, J.H. *Steam Engines*, 2nd ed.; E & F.N. Spon: London, UK, 1890.
13. Novikov, I.I. The efficiency of atomic power stations (A review). *J. Nucl. Energy* **1957**, *7*, 125–128. [CrossRef]
14. Chambadal, P. *Les Centrales Nucleaires*; Armand Colin: Paris, France, 1957; pp. 41–58.
15. Curzon, F.L.; Ahlborn, B. Efficiency of a Carnot engine at maximum power output. *Am. J. Phys.* **1975**, *43*, 22–24. [CrossRef]
16. Hoffman, K.H.; Burzler, J.; Fischer, A.; Schaller, M.; Schubert, S. Optimal process paths for endoreversible systems. *J. Non-Equilib. Thermodyn.* **2003**, *28*, 233–268. [CrossRef]
17. Zaeva, M.A.; Tsirlin, A.M.; Didina, O.V. Finite time thermodynamics: Realizability domain of heat to work converters. *J. Non-Equilib. Thermodyn.* **2019**, *44*, 181–191. [CrossRef]
18. Masser, R.; Hoffmann, K.H. Endoreversible modeling of a hydraulic recuperation system. *Entropy* **2020**, *22*, 383. [CrossRef]
19. Masser, R.; Khodja, A.; Scheunert, M.; Schwalbe, K.; Fischer, A.; Paul, R.; Hoffmann, K.H. Optimized piston motion for an alpha-type Stirling engine. *Entropy* **2020**, *22*, 700. [CrossRef]
20. Muschik, W.; Hoffmann, K.H. Modeling, simulation, and reconstruction of 2-reservoir heat-to-power processes in finite-time thermodynamics. *Entropy* **2020**, *22*, 997. [CrossRef]
21. Andresen, B.; Essex, C. Thermodynamics at very long time and space scales. *Entropy* **2020**, *22*, 1090. [CrossRef]
22. Scheunert, M.; Masser, R.; Khodja, A.; Paul, R.; Schwalbe, K.; Fischer, A.; Hoffmann, K.H. Power-optimized sinusoidal piston motion and its performance gain for an Alpha-type Stirling engine with limited regeneration. *Energies* **2020**, *13*, 4564. [CrossRef]
23. Chen, L.G.; Feng, H.J.; Ge, Y.L. Maximum energy output chemical pump configuration with an infinite-low- and a finite-high-chemical potential mass reservoirs. *Energy Convers. Manag.* **2020**, *223*, 113261. [CrossRef]
24. Qi, C.Z.; Ding, Z.M.; Chen, L.G.; Ge, Y.L.; Feng, H.J. Modeling and performance optimization of an irreversible two-stage combined thermal Brownian heat engine. *Entropy* **2021**, *23*, 419. [CrossRef]
25. Chen, L.G.; Meng, Z.W.; Ge, Y.L.; Wu, F. Performance analysis and optimization for irreversible combined Carnot heat engine working with ideal quantum gases. *Entropy* **2021**, *23*. in press.
26. Bejan, A. Theory of heat transfer-irreversible power plant. *Int. J. Heat Mass Transf.* **1988**, *31*, 1211–1219. [CrossRef]
27. Morisaki, T.; Ikegami, Y. Maximum power of a multistage Rankine cycle in low-grade thermal energy conversion. *Appl. Therm. Eng.* **2014**, *69*, 78–85. [CrossRef]
28. Sadatsakkak, S.A.; Ahmadi, M.H.; Ahmadi, M.A. Thermodynamic and thermo-economic analysis and optimization of an irreversible regenerative closed Brayton cycle. *Energy Convers. Manag.* **2015**, *94*, 124–129. [CrossRef]
29. Yasunaga, T.; Ikegami, Y. Application of finite time thermodynamics for evaluation method of heat engines. *Energy Proc.* **2017**, *129*, 995–1001. [CrossRef]
30. Yasunaga, T.; Fontaine, K.; Morisaki, T.; Ikegami, Y. Performance evaluation of heat exchangers for application to ocean thermal energy conversion system. *Ocean Therm. Energy Convers.* **2017**, *22*, 65–75.
31. Yasunaga, T.; Noguchi, T.; Morisaki, T.; Ikegami, Y. Basic heat exchanger performance evaluation method on OTEC. *J. Mar. Sci. Eng.* **2018**, *6*, 32. [CrossRef]
32. Fontaine, K.; Yasunaga, T.; Ikegami, Y. OTEC maximum net power output using Carnot cycle and application to simplify heat exchanger selection. *Entropy* **2019**, *21*, 1143. [CrossRef]
33. Fawal, S.; Kodal, A. Comparative performance analysis of various optimization functions for an irreversible Brayton cycle applicable to turbojet engines. *Energy Convers. Manag.* **2019**, *199*, 111976. [CrossRef]
34. Yasunaga, T.; Ikegami, Y. Finite-time thermodynamic model for evaluating heat engines in ocean thermal energy conversion. *Entropy* **2020**, *22*, 211. [CrossRef]
35. Yu, X.F.; Wang, C.; Yu, D.R. Minimization of entropy generation of a closed Brayton cycle based precooling-compression system for advanced hypersonic airbreathing engine. *Energy Convers. Manag.* **2020**, *209*, 112548. [CrossRef]
36. Arora, R.; Arora, R. Thermodynamic optimization of an irreversible regenerated Brayton heat engine using modified ecological criteria. *J. Therm. Eng.* **2020**, *6*, 28–42. [CrossRef]
37. Liu, H.T.; Zhai, R.R.; Patchigolla, K.; Turner, P.; Yang, Y.P. Analysis of integration method in multi-heat-source power generation systems based on finite-time thermodynamics. *Energy Convers. Manag.* **2020**, *220*, 113069. [CrossRef]

38. Gonca, G.; Sahin, B.; Cakir, M. Performance assessment of a modified power generating cycle based on effective ecological power density and performance coefficient. *Int. J. Exergy* **2020**, *33*, 153–164. [CrossRef]
39. Karakurt, A.S.; Bashan, V.; Ust, Y. Comparative maximum power density analysis of a supercritical CO_2 Brayton power cycle. *J. Therm. Eng.* **2020**, *6*, 50–57. [CrossRef]
40. Ahmadi, M.H.; Dehghani, S.; Mohammadi, A.H.; Feidt, M.; Barranco-Jimenez, M.A. Optimal design of a solar driven heat engine based on thermal and thermo-economic criteria. *Energy Convers. Manag.* **2013**, *75*, 635–642. [CrossRef]
41. Ahmadi, M.H.; Mohammadi, A.H.; Dehghani, S.; Barranco-Jimenez, M.A. Multi-objective thermodynamic-based optimization of output power of Solar Dish-Stirling engine by implementing an evolutionary algorithm. *Energy Convers. Manag.* **2013**, *75*, 438–445. [CrossRef]
42. Ahmadi, M.H.; Ahmadi, M.A.; Mohammadi, A.H.; Feidt, M.; Pourkiaei, S.M. Multi-objective optimization of an irreversible Stirling cryogenic refrigerator cycle. *Energy Convers. Manag.* **2014**, *82*, 351–360. [CrossRef]
43. Ahmadi, M.H.; Ahmadi, M.A.; Mehrpooya, M.; Hosseinzade, H.; Feidt, M. Thermodynamic and thermo-economic analysis and optimization of performance of irreversible four-temperature-level absorption refrigeration. *Energy Convers. Manag.* **2014**, *88*, 1051–1059. [CrossRef]
44. Ahmadi, M.H.; Ahmadi, M.A. Thermodynamic analysis and optimization of an irreversible Ericsson cryogenic refrigerator cycle. *Energy Convers. Manag.* **2015**, *89*, 147–155. [CrossRef]
45. Jokar, M.A.; Ahmadi, M.H.; Sharifpur, M.; Meyer, J.P.; Pourfayaz, F.; Ming, T.Z. Thermodynamic evaluation and multi-objective optimization of molten carbonate fuel cell-supercritical CO_2 Brayton cycle hybrid system. *Energy Convers. Manag.* **2017**, *153*, 538–556. [CrossRef]
46. Han, Z.H.; Mei, Z.K.; Li, P. Multi-objective optimization and sensitivity analysis of an organic Rankine cycle coupled with a one-dimensional radial-inflow turbine efficiency prediction model. *Energy Convers. Manag.* **2018**, *166*, 37–47. [CrossRef]
47. Ghasemkhani, A.; Farahat, S.; Naserian, M.M. Multi-objective optimization and decision making of endoreversible combined cycles with consideration of different heat exchangers by finite time thermodynamics. *Energy Convers. Manag.* **2018**, *171*, 1052–1062. [CrossRef]
48. Ahmadi, M.H.; Jokar, M.A.; Ming, T.Z.; Feidt, M.; Pourfayaz, F.; Astaraei, F.R. Multi-objective performance optimization of irreversible molten carbonate fuel cell–Braysson heat engine and thermodynamic analysis with ecological objective approach. *Energy* **2018**, *144*, 707–722. [CrossRef]
49. Tierney, M. Minimum exergy destruction from endoreversible and finite-time thermodynamics machines and their concomitant indirect energy. *Energy* **2020**, *197*, 117184. [CrossRef]
50. Yang, H.; Yang, C.X. Derivation and comparison of thermodynamic characteristics of endoreversible aircraft environmental control systems. *Appl. Therm. Eng.* **2020**, *180*, 115811. [CrossRef]
51. Tang, C.Q.; Chen, L.G.; Feng, H.J.; Ge, Y.L. Four-objective optimization for an irreversible closed modified simple Brayton cycle. *Entropy* **2021**, *23*, 282. [CrossRef] [PubMed]
52. Lichty, C. *Combustion Engine Processes*; McGraw-Hill: New York, NY, USA, 1967.
53. Georgiou, D.P. Useful work and the thermal efficiency in the ideal Lenoir with regenerative preheating. *J. Appl. Phys.* **2008**, *88*, 5981–5986. [CrossRef]
54. Shen, X.; Chen, L.G.; Ge, Y.L.; Sun, F.R. Finite-time thermodynamic analysis for endoreversible Lenoir cycle coupled to constant-temperature heat reservoirs. *Int. J. Energy Environ.* **2017**, *8*, 272–278.
55. Ahmadi, M.H.; Nazari, M.A.; Feidt, M. Thermodynamic analysis and multi-objective optimisation of endoreversible Lenoir heat engine cycle based on the thermo-economic performance criterion. *Int. J. Ambient Energy* **2019**, *40*, 600–609. [CrossRef]

Article

Four-Objective Optimizations for an Improved Irreversible Closed Modified Simple Brayton Cycle

Chenqi Tang [1,2,3], Lingen Chen [1,2,*], Huijun Feng [1,2,*] and Yanlin Ge [1,2]

[1] Institute of Thermal Science and Power Engineering, Wuhan Institute of Technology, Wuhan 430205, China; tangchenqi7@163.com (C.T.); geyali9@hotmail.com (Y.G.)
[2] School of Mechanical & Electrical Engineering, Wuhan Institute of Technology, Wuhan 430205, China
[3] College of Power Engineering, Naval University of Engineering, Wuhan 430033, China
[*] Correspondence: lgchenna@yahoo.com or lingenchen@hotmail.com (L.C.); huijunfeng@139.com or 1507136871@139.com (H.F.); Tel.: +86-27-836150466 (L.C.); Fax: +86-27-83638709 (L.C.)

Abstract: An improved irreversible closed modified simple Brayton cycle model with one isothermal heating process is established in this paper by using finite time thermodynamics. The heat reservoirs are variable-temperature ones. The irreversible losses in the compressor, turbine, and heat exchangers are considered. Firstly, the cycle performance is optimized by taking four performance indicators, including the dimensionless power output, thermal efficiency, dimensionless power density, and dimensionless ecological function, as the optimization objectives. The impacts of the irreversible losses on the optimization results are analyzed. The results indicate that four objective functions increase as the compressor and turbine efficiencies increase. The influences of the latter efficiency on the cycle performances are more significant than those of the former efficiency. Then, the NSGA-II algorithm is applied for multi-objective optimization, and three different decision methods are used to select the optimal solution from the Pareto frontier. The results show that the dimensionless power density and dimensionless ecological function compromise dimensionless power output and thermal efficiency. The corresponding deviation index of the Shannon Entropy method is equal to the corresponding deviation index of the maximum ecological function.

Keywords: closed simple Brayton cycle; power output; thermal efficiency; power density; ecological function; multi-objective optimization

Citation: Tang, C.; Chen, L.; Feng, H.; Ge, Y. Four-Objective Optimizations for an Improved Irreversible Closed Modified Simple Brayton Cycle. *Entropy* **2021**, *23*, 282. https://doi.org/10.3390/e23030282

Academic Editor: Michel Feidt

Received: 17 January 2021
Accepted: 22 February 2021
Published: 26 February 2021

Publisher's Note: MDPI stays neutral with regard to jurisdictional claims in published maps and institutional affiliations.

1. Introduction

Some scholars have studied performances of gas turbine plants (Brayton cycle (BCY)) [1–4] all over the world for their small size and comprehensive energy sources. The gas-steam combined, cogeneration, and other complex cycles have appeared for the requirements of energy conservation and environmental protection. The thermal efficiency (η) of a simple BCY is low, and the NOx content in combustion product is high. To further improve the cycle performance, it has become a key research direction to improve the initial temperature of the gas or to adopt the advanced cycles (such as regenerative, intercooled, intercooled and regenerative, isothermal heating, and other complex combined cycles).

In the case of simple heating, when the compressible subsonic gas flows through the smooth heating pipe with the fixed cross-sectional area, the gas temperature increases along the pipe direction; in the case of simple region change, when the compressible subsonic gas flows through the smooth adiabatic reductive pipe, the gas temperature decreases along the pipe direction. Based on these two gas properties, the isothermal heating process (IHP) can be realized when the compressible subsonic gas flows through the smooth heating reductive pipe. The combustion chamber, which can recognize the IHP, is called the convergent combustion chamber (CCC). The pipe of the CCC is assumed to be smooth. During the heating process, the temperature of the gas is always constant. According to the energy conservation law, the kinetic energy of the gas increases, that is, the pushing work of

Entropy **2021**, *23*, 282. https://doi.org/10.3390/e23030282 https://www.mdpi.com/journal/entropy

the gas increases. From the definition of enthalpy, it can be seen that enthalpy includes two parts: the thermodynamic energy and the pushing work. Therefore, the enthalpy increases. Based on this, Vecchiarelli et al. [5] proposed the CCC to perform the IHP of the working fluid. The power output ($\overline{W}$) and η of the BCY could be improved, and the emission of harmful gases such as NOx could be reduced by adding this combustion chamber model. The regenerative BCYs [6–8] and binary BCY [9] with IHPs were also studied by applying the classical thermodynamics.

Finite time thermodynamics (FTT) is a useful thermodynamic analysis theory and method [10–19]. In general, it is known that Curzon and Ahlborn [12] initialized FTT in 1975. In fact, the classical efficiency bound at the maximum power was also derived by Moutier [10] in 1872 and Novikov [11] in 1957. The applications of FTT include majorly two fields: optimal configurations [20–36] and optimal performances [37–61] studies for thermodynamic cycles and processes. The W and η have been often considered as the optimization objectives (OPOs) of the heat engines [62–72]. When the power density (P) [73–81] was taken as the OPO, the operating unit had a smaller size and higher η. Aditionally, the ecological function (E) [82–88] is also an OPO that balances the conflict between W and η.

Kaushik et al. [89] first applied the FTT to studying the regenerative BCY with an IHP. The regenerative, intercooled and regenerative complex BCYs with isothermal heating combustor were further investigated [90–96]. Based on this, Chen et al. [97–99] studied the endoreversible simple isothermal heating BCY with the W, η and E as OPOs. Arora et al. [100,101] adopted NSGA-II and evolutionary algorithms to optimize the irreversible isothermal heating regenerative BCY with the W and η as the OPOs. Chen et al. [102] considered the variable isothermal pressure drop ratio (π_t), established an improved isothermal heating regenerative BCY model, and studied the regenerator's role on cycle performance. Qi et al. [103] demonstrated a closed endoreversible modified binary BCY with IHPs and found the W and η raised as the heat reservoirs' temperature ratios. Tang et al. [104] considered the variable π_t and established an improved irreversible binary BCY model modified by isothermal heating. The heat exchanger's heat conductance distributions (HCDs) and the top and bottom cycles' pressure ratios were taken as optimization variables to optimize the cycle performance.

In the process of the thermodynamic system optimization, single-objective optimization often led to unacceptable objectives for other objectives when there were conflicts among the considered goals. Multi-objective optimization would consider the trade-offs among the goals, and the optimized results were more reasonable [99,100,102,105–125].

In applying the FTT, the heat transfer was introduced into the thermodynamic analysis of the thermodynamic process, and finite temperature difference was considered in Refs. [11,12]. In this paper, the same method in Refs. [11,12] will be used, and the finite temperature difference will be considered when establishing the model, which is the key relation among this paper and the Refs. [11,12]. On this basis, the cycle's irreversibility will be further considered, and the corresponding conclusion will be more in line with the actual situation. The compression and expansion losses in the model in Refs. [97–99] were not considered, and they will be further considered in this paper alongside the losses in the heat exchangers. Meanwhile, the thermal resistance loss and the optimal HCD will be considered. With the W, η, P and E, respectively, as the OPOs, an improved irreversible closed modified simple BCY with one IHP and coupled to variable-temperature heat reservoirs (VTHRs) will be optimized, and the optimization results will be compared. The effects of the compressor and turbine efficiencies on optimization results will be analyzed. The NSGA-II algorithm will be applied for multi-objective optimization to obtain the Pareto frontier further. The results obtained in this paper will reveal the original results in Refs. [10–12], which were the initial work of the FTT theory.

2. Cycle Model and Performance Analytical Indicators

The schematic diagram of an improved irreversible closed modified simple BCY with one IHP and coupled to VTHRs is shown in Figure 1. A compressor (C), a regular combustion chamber (RCC), a CCC, a turbine (T), and a precooler are the main parts of the cycle. The corresponding $T - s$ diagram of the cycle is shown in Figure 2. The cycle consists of five processes in total:

1. The process $1 \rightarrow 2$ is an irreversible adiabatic compression process in C, and the process $1 \rightarrow 2s$ is an isentropic process corresponding to the process $1 \rightarrow 2$.
2. The process $2 \rightarrow 3$ is an isobaric endothermic process in RCC.
3. The process $3 \rightarrow 4$ is an IHP in CCC. In CCC, the working fluid is isothermally heated, and its flow velocity rises from V_3 to V_4 (the Mach number increases from M_3 to M_4), and its specific enthalpy rises from h_3 to h_4. The parameter $\pi_t (= p_4/p_3 \leq 1)$ is the isothermal pressure drop ratio. The π_t needs to be given in Refs. [97,98], but the π_t of the improved cycle established in this paper will change with the operation state. The degree of the IHP can be represented by π_t, and the greater the π_t, the greater the degree.
4. The process $4 \rightarrow 5$ is an adiabatic exothermic process in turbine, and the process $4 \rightarrow 5s$ is the isentropic process corresponding to the process $4 \rightarrow 5$.
5. The process $5 \rightarrow 1$ is an isobaric exothermic process in a precooler.

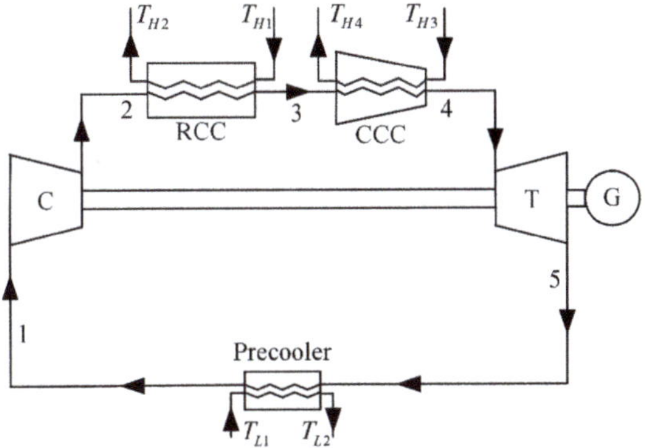

Figure 1. Schematic diagram of the cycle.

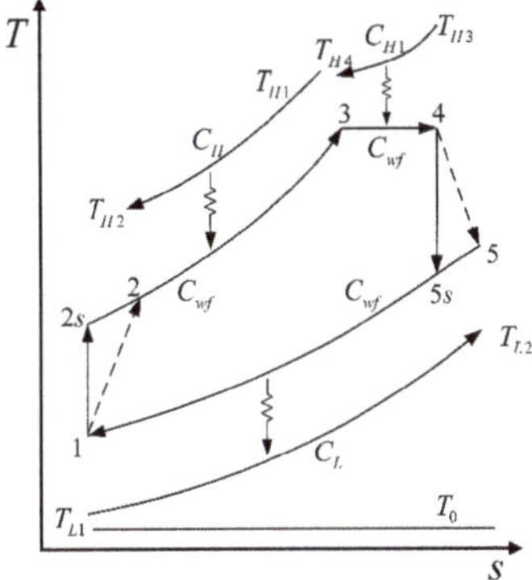

Figure 2. Diagram of the cycle.

The working fluid is the ideal gas. The pressures and temperatures of the working fluid are $p_i (i = 1,\ 2,\ 3,\ 4,\ 5,\ 2s,\ 5s)$ and T_i, and the ratio of specific heat is k. The outside fluids' temperatures are $T_j (j = H1,\ H2,\ H3,\ H4,\ L1,\ L2)$. The specific heat at constant pressure and the working fluid's mass flow rate are C_p and $\dot{m}$. The working fluid's thermal capacity rate is C_{wf} where $C_{wf} = C_p \dot{m}$. The outer fluids' thermal capacity rates at the RCC, CCC, and precooler are C_H, C_{H1} and C_L, respectively; then, one has:

$$C_{Hmax} = \max\{C_H,\ C_{wf}\},\ C_{Lmax} = \max\{C_L,\ C_{wf}\},\ C_{Hmin} = \min\{C_H,\ C_{wf}\},\ C_{Lmin} = \min\{C_L,\ C_{wf}\} \tag{1}$$

The heat exchangers' heat conductance is the product of the heat transfer coefficient and the heat transfer area. The heat exchangers' heat conductance values in the RCC, CCC, and precooler are U_H, U_{H1} and U_L, the heat transfer units' numbers are N_H, N_{H1} and N_L, and the effectiveness values are E_H, E_{H1} and E_L, respectively:

$$N_H = U_H / C_{Hmin},\ N_{H1} = U_{H1} / C_{H1},\ N_L = U_L / C_{Lmin} \tag{2}$$

$$E_H = \frac{1 - e^{-N_H(1 - C_{Hmin}/C_{Hmax})}}{1 - (C_{Hmin}/C_{Hmax})e^{-N_H(1 - C_{Hmin}/C_{Hmax})}} \tag{3}$$

$$E_{H1} = 1 - e^{-N_{H1}} \tag{4}$$

$$E_L = \frac{1 - e^{-N_L(1 - C_{Lmin}/C_{Lmax})}}{1 - (C_{Lmin}/C_{Lmax})e^{-N_L(1 - C_{Lmin}/C_{Lmax})}} \tag{5}$$

When $C_{Hmax} = C_{Hmin}$ and $C_{Lmax} = C_{Lmin}$, Equations (3) and (5) are, respectively, simplified as:

$$E_H = N_H / (N_H + 1) \tag{6}$$

$$E_L = N_L / (N_L + 1) \tag{7}$$

The outside fluids' temperature ratios at the RCC and CCC are:

$$\tau_{H1} = T_{H1} / T_0 \tag{8}$$

$$\tau_{H3} = T_{H3} / T_0 \tag{9}$$

where T_0 is the ambient temperature.

The process $1 \rightarrow 2s$ is the isentropic one, namely:

$$T_{2s} / T_1 = \pi^m = x \tag{10}$$

where $m = (k - 1)/k$ and π is the pressure ratio of the compressor.

The process $4 \rightarrow 5s$ is the isentropic one, namely:

$$T_4 / T_{5s} = \pi^m \pi_t^m = xy \tag{11}$$

The process $3 \rightarrow 4$ is the isothermal one, namely:

$$T_3 = T_4 \tag{12}$$

$$\dot{Q}_{3-4} = \dot{m}(h_4 - h_3) - \dot{m}\int_3^4 vdp = -\dot{m}R_g T_3 \ln \pi_t \tag{13}$$

where π_t, M_3 and M_4 must satisfy the following relation:

$$\ln \pi_t = -c_p(k - 1)(M_4^2 - M_3^2)/(2R_g) \tag{14}$$

where the working fluid's flow velocity must be subsonic, namely, M_3, $M_4 < 1$ Because the working fluid has an initial speed, $(M_4^2 - M_3^2) < 0.96$ and $\pi_t > 0.5107$ when $M_3 = 0.2$.

Because of $M_4 > M_3$, $\pi_t < 1$. When $\pi_t = 1$, the cycle model in this paper can be simplified to a simple Brayton cycle.

According to the definition of π_t, it can be obtained that:

$$\pi_t = \frac{p_4}{p_3} = \frac{p_4}{p_3} \cdot \frac{p_1}{p_1} = \frac{p_4}{p_1} \cdot \pi^{-1} \geq \pi^{-1} \tag{15}$$

Considering the irreversibilities in the compressor and the turbine, the efficiencies of them are:

$$\eta_c = (T_1 - T_{2s})/(T_1 - T_2) \tag{16}$$

$$\eta_t = (T_5 - T_4)/(T_{5s} - T_4) \tag{17}$$

The pressure drop is not considered in this paper. It will be considered in future, as it was by Ref. [126]. The study in Ref. [126] showed that the pressure drop loss has a little influence on the cycle performance quantitatively, and has no influence qualitatively.

The working fluid's heat absorption rates at RCC and CCC are $\dot{Q}_{2-3}$ and $\dot{Q}_{3-4}$, respectively:

$$\dot{Q}_{2-3} = C_H(T_{H1} - T_{H2}) = C_{wf}(T_3 - T_2) = C_{Hmin}E_H(T_{H1} - T_2) \tag{18}$$

$$\dot{Q}_{3-4} = C_{H1}(T_{H3} - T_{H4}) = C_{H1}E_{H1}(T_{H3} - T_3) = \dot{m}(V_4^2 - V_3^2)/2 \tag{19}$$

The heat releasing rate at the precooler is $\dot{Q}_{5-1}$, namely:

$$\dot{Q}_{5-1} = C_L(T_{L2} - T_{L1}) = C_{wf}(T_5 - T_1) = C_{Lmin}E_L(T_5 - T_{L1}) \tag{20}$$

The heat leakages between the heat source and the environment [127,128] are neglected. Therefore, the W and η are:

$$W = \dot{Q}_{2-3} + \dot{Q}_{3-4} - \dot{Q}_{5-1} \tag{21}$$

$$\eta = W/(\dot{Q}_{2-3} + \dot{Q}_{3-4}) \tag{22}$$

The dimensionless power output ($\overline{W}$) is:

$$\overline{W} = W/(C_{wf}T_0) \tag{23}$$

The maximum specific volume corresponding to state point 5 is v_5. The P is calculated as:

$$P = W/v_5 \tag{24}$$

The specific volume corresponding to state point 1 is v_1. The dimensionless power density ($\overline{P}$) and dimensionless maximum specific volume (v_5/v_1) are obtained as:

$$\overline{P} = \frac{P}{C_{wf}T_0/v_1} = \frac{W/v_5}{C_{wf}T_0/v_1} = \frac{W}{C_{wf}T_0} \times \frac{T_1}{T_5} = \overline{W} \times \frac{T_1}{T_5} \tag{25}$$

$$v_5/v_1 = T_5/T_1 \tag{26}$$

There are two different methods for calculating the entropy production rate. One was suggested by Bejan [129,130], and the another was suggested by Salamon et al. [131]. In this article, the method used is the one suggested by the latter.

The entropy production rate (s_g) and E are, respectively, calculated as:

$$s_g = C_H \ln(T_{H2}/T_{H1}) + C_{H1} \ln(T_{H4}/T_{H3}) + C_L \ln(T_{L2}/T_{L1}) \tag{27}$$

$$E = W - T_0 s_g \tag{28}$$

The dimensionless ecological function ($\overline{E}$) is obtained as:

$$\overline{E} = E/(C_{wf}T_0) \tag{29}$$

Equations (10)–(12) and (16)–(29) are combined, and the four dimensionless performance indicators of the cycle are obtained as follows:

$$\overline{W} = \frac{\begin{aligned}&C_{wf}xy(C_{H1}E_{H1}T_{H3}+C_{Lmin}E_LT_{L1})+C_{Hmin}E_HT_{H1}\big\{xy[C_{wf}\\&-C_{H1}E_{H1}+C_{Lmin}E_L(\eta_t-1)]-C_{Lmin}E_L\eta_t\big\}+a_1\big\{C_{Lmin}E_L\\&\times[(\eta_t-1)xy-\eta_t](C_{wf}-E_HC_{Hmin})-xy[C_{wf}C_{Hmin}E_H\\&+C_{H1}E_{H1}(C_{wf}-C_{Hmin}E_H)]\big\}\end{aligned}}{C_{wf}^2 T_0 xy} \tag{30}$$

$$\eta = \frac{\begin{aligned}&C_{Hmin}C_{Lmin}E_HE_L\eta_tT_{H1}-\big\{C_{Hmin}E_HT_{H1}[C_{wf}-C_{H1}E_{H1}+C_{Lmin}E_L(\eta_t-1)]\\&+C_{wf}xy(C_{H1}E_{H1}T_{H3}+C_{Lmin}E_LT_{L1})\big\}+a_1\big\{[C_{Hmin}C_{wf}E_H+C_{H1}E_{H1}(C_{wf}\\&-E_HC_{Hmin})]xy-C_{Lmin}E_L(C_{wf}-C_{Hmin}E_H)[(\eta_t-1)xy-\eta_t]\big\}\end{aligned}}{\begin{aligned}&xy\big\{a_1[C_{H1}C_{wf}E_{H1}+C_{Hmin}E_H(C_{wf}-C_{H1}E_{H1})]+C_{Hmin}E_H(C_{H1}E_{H1}\\&-C_{wf})T_{H1}-C_{H1}C_{wf}E_{H1}T_{H3}\big\}\end{aligned}} \tag{31}$$

$$\overline{P} = \frac{\begin{aligned}&\big\{a_1(C_{wf}-C_{Hmin}E_H)(C_{wf}-C_{Lmin}E_L)[xy(\eta_t-1)-\eta_t]-C_{Lmin}C_{wf}E_LT_{L1}x\\&\times y+E_HC_{Hmin}T_{H1}(C_{wf}-C_{Lmin}E_L)[(\eta_t-1)xy-\eta_t]\big\}\big\{C_{wf}xy(C_{H1}E_{H1}T_{H3}\\&+C_{Lmin}E_LT_{L1})+\big\{xy[C_{wf}-C_{H1}E_{H1}+C_{Lmin}E_L(\eta_t-1)]-C_{Lmin}E_L\eta_t\big\}C_{Hmin}\\&\times E_HT_{H1}+a_1\big\{C_{Lmin}(C_{wf}-E_HC_{Hmin})E_L\big\{(\eta_t-1)xy-\eta_t-xy[C_{Hmin}C_{wf}E_H\\&+C_{H1}E_{H1}(C_{wf}-C_{Hmin}E_H)]\big\}\big\}\big\}\end{aligned}}{C_{wf}^3 T_0 xy[a_1(C_{wf}-C_{Hmin}E_H)+C_{Hmin}E_HT_{H1}][(\eta_t-1)xy-\eta_t]} \tag{32}$$

$$\overline{E} = \frac{\begin{aligned}&\big\{C_{wf}xy(C_{H1}E_{H1}T_{H3}+C_{Lmin}E_LT_{L1})+C_{Hmin}E_HT_{H1}\big\{xy[C_{wf}-C_{H1}E_{H1}\\&+C_{Lmin}E_L(\eta_t-1)]-C_{Lmin}E_L\eta_t\big\}+a_1\big\{C_{Lmin}E_L(C_{wf}-C_{Hmin}E_H)[(\eta_t\\&-1)xy-\eta_t]-xy[C_{wf}C_{Hmin}E_H+C_{H1}E_{H1}(C_{wf}-C_{Hmin}E_H)]\big\}\big\}/(T_0\\&\times xy)-C_{wf}\big\{C_L\ln\big\{1+\big\{C_{Lmin}E_L\big\{a_1C_{wf}\eta_t-C_{wf}xy[a_1(\eta_t-1)+T_{L1}]\\&+C_{Hmin}E_H(a_1-T_{H1})[(\eta_t-1)xy-\eta_t]\big\}\big\}/(C_LC_{wf}T_{L1}xy)\big\}+C_H\ln\{[a_1\\&\times C_{Hmin}E_H+(C_H-C_{Hmin}E_H)T_{H1}]/(C_HT_{H1})\}+C_{H1}\ln\big\{1+\big\{E_{H1}[C_{wf}\\&\times(a_1-T_{H3})+E_HC_{Hmin}(T_{H1}-a_1)]\big\}/(C_{wf}T_{H3})\big\}\big\}\end{aligned}}{C_{wf}^2} \tag{33}$$

where

$$a_1 = \frac{(\eta_c+x-1)\big\{C_{Lmin}C_{wf}E_LT_{L1}xy-C_{Hmin}E_HT_{H1}(C_{wf}-C_{Lmin}E_L)[(\eta_t-1)xy-\eta_t]\big\}}{\begin{aligned}&C_{Hmin}C_{Lmin}E_HE_L(\eta_c+x-1)(\eta_txy-xy-\eta_t)+C_{wf}^2[xy-x^2y+\eta_t(\eta_c+x\\&-1)(xy-1)]-C_{wf}(\eta_c+x-1)(E_HC_{Hmin}+E_LC_{Lmin})[(\eta_t-1)xy-\eta_t]\end{aligned}} \tag{34}$$

Parameters x and y in Equations (30)–(34) can be obtained by Equations (13) and (19), and then the arithmetic solution of $\overline{W}$, η, $\overline{P}$ and $\overline{E}$ can be gained. When C_H, C_{H1}, C_L, E_H, E_{H1}, E_L, η_c and η_t are specific values, the cycle could be transformed into different cycle models. Equations (30)–(34) could be simplified into the performance indicators of the various cycle models, which have certain universality.

1. When $C_{H1} = C_L \to \infty$, Equations (30)–(34) can be simplified into the performance indicators of the irreversible simple BCY with an IHP and coupled to constant-temperature heat reservoirs (CTHRs) whose $T - s$ diagram is shown in Figure 3a:

$$\overline{W} = \frac{\begin{aligned}&C_{wf}xy(C_{H1}E_{H1}T_{H3} + C_{Lmin}E_LT_{L1}) + C_{Hmin}E_HT_{H1}\Big\{xy[C_{wf} \\ &-C_{H1}E_{H1} + C_{Lmin}E_L(\eta_t - 1)] - C_{Lmin}E_L\eta_t\} + a_1\{C_{Lmin}E_L \\ &\times[(\eta_t - 1)xy - \eta_t](C_{wf} - E_HC_{Hmin}) - xy[C_{wf}C_{Hmin}E_H \\ &+C_{H1}E_{H1}(C_{wf} - C_{Hmin}E_H)]\big\}\end{aligned}}{C_{wf}^2 T_0 xy} \tag{35}$$

$$\eta = \frac{\begin{aligned}&C_{wf}E_HE_L\eta_tT_{H1} - \Big\{E_HT_{H1}[C_{wf} - C_{H1}E_{H1} + C_{wf}E_L(\eta_t - 1)] + (C_{H1}E_{H1}T_{H3} \\ &+C_{wf}E_LT_{L1})\Big\}xy + a_2\Big\{[C_{wf}E_H + C_{H1}E_{H1}(1 - E_H)]xy - C_{wf}E_L(1 - E_H) \\ &\times[-\eta_t + (-1 + \eta_t)xy]\Big\}\end{aligned}}{xy\Big\{a_2[C_{H1}E_{H1} + E_H(C_{wf} - C_{H1}E_{H1})] + E_HT_{H1}(C_{H1}E_{H1} - C_{wf}) - C_{H1}E_{H1}T_{H3}\Big\}} \tag{36}$$

$$\overline{P} = \frac{\begin{aligned}&C_{wf}\{-E_LT_{L1}xy + a_2(1 - E_H)(1 - E_L)[(\eta_t - 1)xy - \eta_t] + E_HT_{H1}(1 - E_L)[(\eta_t \\ &-1)xy - \eta_t]\}\Big\{xy(C_{H1}E_{H1}T_{H3} + C_{wf}E_LT_{L1}) + E_HT_{H1}\Big\{xy[C_{wf} - C_{H1}E_{H1} + C_{wf} \\ &\times E_L(\eta_t - 1)] - C_{wf}E_L\eta_t\Big\} + a_2C_{wf}(1 - E_H)E_L\Big\{(\eta_t - 1)xy - \eta_t - C_{wf}xy[C_{wf}E_H \\ &+C_{H1}E_{H1}(1 - E_H)]\}\Big\}\end{aligned}}{C_{wf}^3 T_0 xy[a_2(1 - E_H) + E_HT_{H1}][(\eta_t - 1)xy - \eta_t]} \tag{37}$$

$$\overline{E} = \frac{\begin{aligned}&\Big\{xy(C_{H1}E_{H1}T_{H3} + C_{wf}E_LT_{L1}) + E_HT_{H1}\Big\{[C_{wf} - C_{H1}E_{H1} + C_{wf}E_L(\eta_t - 1)]xy - E_L\eta_t\Big\} \\ &+a_2C_{wf}\Big\{C_{wf}E_L(1 - E_H)[(\eta_t - 1)xy - \eta_t] - xy[C_{wf}E_H + C_{H1}(1 - E_H)E_{H1}]\Big\}\Big\}/(T_0xy) \\ &-\Big\{C_H\ln[(a_2C_{wf}E_H + C_HT_{H1} - C_{wf}E_HT_{H1})/(C_HT_{H1})] + C_{H1}\ln\Big\{1 + \Big\{E_{H1}[a_2 + C_{wf}E_H \\ &\times(T_{H1} - a_2)/C_{wf} - T_{H3}]\Big\}/T_{H3}\Big\} + C_L\ln\Big\{1 + \Big\{C_{wf}E_L\{a_2\eta_t - xy[a_2(\eta_t - 1) + T_{L1}] + E_H \\ &\times(a_2 - T_{H1})[(\eta_t - 1)xy - \eta_t]\}\Big\}/(C_LT_{L1}xy)\Big\}\Big\}\end{aligned}}{C_{wf}} \tag{38}$$

where

$$a_2 = \frac{(\eta_c + x - 1)\{-E_LT_{L1}xy - E_HT_{H1}(1 - E_L)[(\eta_t - 1)xy - \eta_t]\}}{\begin{aligned}&E_HE_L(\eta_c + x - 1)[(xy - 1)\eta_t - xy] + [xy - x^2y + \eta_t(\eta_c + x - 1) \\ &\times(xy - 1)] - [(\eta_t - 1)xy - \eta_t](\eta_c + x - 1)(E_H + E_L)\end{aligned}} \tag{39}$$

2. When $\eta_{c1} = \eta_{t1} = 1$, Equations (30)–(34) can be respectively simplified into the performance indicators of the endoreversible simple BCY with an IHP and coupled to VTHRs [99], whose $T - s$ diagram is shown in Figure 3b:

$$\overline{W} = \frac{\begin{aligned}&C_{wf}x\Big\{C_{Lmin}C_{wf}E_LT_{L1}(y - 1) + C_{H1}E_{H1}[C_{wf}T_{H3}(y - 1) + C_{Lmin}E_L(T_{H3} \\ &-T_{L1}xy)]\Big\} + E_HC_{Hmin}\Big\{C_{Lmin}E_L[C_{wf}T_{H1}(x - 1) + C_{wf}T_{L1}x(1 - xy) + C_{H1} \\ &\times E_{H1}x(T_{L1}xy - T_{H3})] + xC_{wf}[(y - 1)C_{wf}T_{H1} + C_{H1}E_{H1}(T_{H3} - T_{H1}y)]\Big\}\end{aligned}}{C_{wf}T_0x[C_{wf}^2y - (C_{wf} - C_{Hmin}E_H)(C_{wf} - C_{Lmin}E_L)]} \tag{40}$$

$$\eta = \frac{\begin{aligned}&C_{wf}T_0x\Big\{C_{Lmin}C_{wf}E_LT_{L1}(y - 1) + C_{H1}E_{H1}[C_{wf}T_{H3}(y - 1) + C_{Lmin}E_L(T_{H3} - T_{L1}xy)]\Big\} \\ &+C_{Hmin}E_H\Big\{C_{Lmin}E_L[C_{wf}T_{H1}(x - 1) + C_{wf}T_{L1}x(1 - xy) + C_{H1}E_{H1}x(T_{L1}xy - T_{H3})] \\ &+C_{wf}x[C_{wf}T_{H1}(y - 1) + C_{H1}E_{H1}(T_{H3} - T_{H1}y)]\Big\}\end{aligned}}{C_{wf}T_0x\{C_{Hmin}E_H[C_{wf}^2T_{H1}(y - 1) + C_{H1}C_{wf}E_{H1}(T_{H3} - T_{H1}y) + C_{Lmin}C_{wf}E_L(T_{H1} - T_{L1}xy) + C_{H1}C_{Lmin}E_{H1}E_L(T_{L1}xy - T_{H3})] + C_{H1}C_{wf}E_{H1}[C_{wf}T_{H3}(y - 1) + C_{Lmin}E_L(T_{H3} - T_{L1}xy)]\}} \tag{41}$$

$$\bar{P} = \frac{[C_{Hmin}E_H T_{H1}(C_{wf} - C_{Lmin}E_L) + C_{Lmin}C_{wf}E_L T_{L1}xy]\left\{C_{wf}x\left\{C_{Lmin}C_{wf}E_L T_{L1}(y-1) + C_{H1}E_{H1}\times[C_{wf}T_{H3}(y-1) + C_{Lmin}E_L(T_{H3} - T_{L1}xy)]\right\} + C_{Hmin}E_H\left\{C_{wf}x[C_{wf}T_{H1}(y-1) + (T_{H3} - T_{H1}\times y)C_{H1}E_{H1}] + C_{Lmin}E_L[C_{wf}T_{H1}(x-1) + C_{wf}T_{L1}x(1-xy) + C_{H1}E_{H1}x(T_{L1}xy - T_{H3})]\right\}\right\}}{C_{wf}T_0 x[C_{wf}^2 y - (C_{wf} - C_{Hmin}E_H)(C_{wf} - C_{Lmin}E_L)][C_{Lmin}(C_{wf} - C_{Hmin}E_H)E_L \times T_{L1}x + C_{Hmin}C_{wf}E_H T_{H1}]} \tag{42}$$

$$\bar{E} = \frac{C_{wf}x\left\{C_{Lmin}C_{wf}E_L T_{L1}(y-1) + C_{H1}E_{H1}[C_{wf}T_{H3}(y-1) + C_{Lmin}E_L(T_{H3} - T_{L1}xy)]\right\} + C_{Hmin}E_H\left\{C_{Lmin}E_L[C_{wf}T_{H1}(x-1) + C_{wf}T_{L1}x(1-xy) + C_{H1}\times E_{H1}x(T_{L1}xy - T_{H3})] + C_{wf}x[C_{wf}T_{H1}(y-1) + (T_{H3} - T_{H1}y)C_{H1}E_{H1}]\right\}}{C_{wf}T_0 x[C_{wf}^2 y - (C_{wf} - C_{Hmin}E_H)(C_{wf} - C_{Lmin}E_L)]} \tag{43}$$

$$-\frac{C_H}{C_{wf}T_0}\ln\left\{1 + \frac{C_{Hmin}C_{wf}E_H(C_{wf}T_{H1} - C_{Lmin}E_L T_{H1} - C_{wf}T_{H1}y + C_{Lmin}\times E_L T_{L1}xy)}{T_{H1}[C_H C_{wf}^2 y - C_H(C_{wf} - C_{Hmin}E_H)(C_{wf} - C_{Lmin}E_L)]}\right\}$$

$$-\frac{C_{H1}}{C_{wf}T_0}\ln\frac{\left\{(C_{wf} - C_{Hmin}E_H)(E_{H1} - 1)(C_{wf} - C_{Lmin}E_L)T_{H3} + C_{wf}y\times[C_{Hmin}E_H E_{H1}T_{H1} - C_{wf}(E_{H1}-1)T_{H3}] + C_{Lmin}E_{H1}E_L T_{L1}xy\times(C_{wf} - C_{Hmin}E_H)\right\}}{C_{wf}^2 T_{H3}y - T_{H3}(C_{wf} - C_{Hmin}E_H)(C_{wf} - C_{Lmin}E_L)}$$

$$-\frac{C_L}{C_{wf}T_0}\ln\left\{1 + \frac{C_{Lmin}C_{wf}E_L[C_{Hmin}E_H(T_{H1} - T_{L1}x) - C_{wf}T_{L1}x(y-1)]}{C_L T_{L1}[(C_{Hmin}E_H - C_{wf})(C_{wf}C_{Lmin}E_L)x + C_{wf}^2 xy]}\right\}$$

3. When $\eta_{c1} = \eta_{t1} = 1$ and $C_{H1} = C_{H2} = C_L \to \infty$, Equations (30)–(34) can be simplified into the performance indicators of the endoreversible simple BCY with an IHP and coupled to CTHRs, whose $T - s$ diagram is shown in Figure 3c:

$$\bar{W} = \frac{C_{wf}x\left\{C_{wf}E_L T_{L1}(y-1) + C_{H1}E_{H1}[E_L T_{H3} - E_L T_{L1}xy + T_{H3}(y-1)]\right\} + C_{wf}E_H\left\{E_L[T_{H1}C_{wf}(x-1) + C_{wf}T_{L1}x(1-xy) + C_{H1}E_{H1}x(T_{L1}xy - T_{H3})] + x[C_{wf}T_{H1}(y-1) + C_{H1}E_{H1}(T_{H3} - T_{H1}y)]\right\}}{C_{wf}^2 T_0 x(E_H + E_L + y - E_H E_L - 1)} \tag{44}$$

$$\eta = \frac{T_0 x\left\{C_{wf}E_L T_{L1}y - C_{wf}E_L T_{L1} + C_{H1}E_{H1}[T_{H3}y - T_{H3} + T_{H3}E_L - E_L T_{L1}xy]\right\} + \left\{E_H\left\{x[C_{wf}T_{H1}y - C_{wf}T_{H1} + C_{H1}E_{H1}(T_{H3} - T_{H1}y)] + E_L[C_{wf}T_{H1}x - C_{wf}\times T_{H1} + C_{wf}T_{L1}x(1-xy) + C_{H1}E_{H1}x(-T_{H3} + T_{L1}xy)]\right\}\right\}}{C_{wf}T_0 x\left\{[C_{wf}T_{H1}y - C_{wf}T_{H1} + C_{H1}E_{H1}(T_{H3} - T_{H1}y) + C_{wf}E_L(T_{H1} - T_{L1}xy) + C_{H1}E_{H1}E_L(T_{L1}xy - T_{H3})]E_H + C_{H1}E_{H1}[T_{H3}(y-1) + E_L(T_{H3} - T_{L1}xy)]\right\}} \tag{45}$$

$$\bar{P} = \frac{[E_H T_{H1}(1 - E_L) + E_L T_{L1}xy]\left\{C_{wf}x\left\{C_{wf}E_L T_{L1}(y-1) + C_{H1}E_{H1}[T_{H3}(y-1) + E_L(T_{H3} - T_{L1}xy)]\right\} + E_H C_{wf}\left\{x[C_{wf}T_{H1}(y-1) + C_{H1}E_{H1}(T_{H3} - T_{H1}y)] + E_L[C_{wf}T_{H1}(x-1) + C_{wf}T_{L1}x(1-xy) + C_{H1}xE_{H1}(T_{L1}xy - T_{H3})]\right\}\right\}}{C_{wf}^2 T_0 x(E_H + E_L + y - E_H E_L - 1)(E_H T_{H1} + E_L T_{L1}x - E_L T_{L1}xE_H)} \tag{46}$$

$$
\bar{E} = \dfrac{\begin{aligned}&\left\{C_{wf}E_LT_{L1}(y-1)+C_{H1}E_{H1}[T_{H3}(y-1)+E_LT_{H3}T_{L1}xy]\right\}x\\&+E_H\left\{E_L[C_{wf}T_{H1}(x-1)+(1-xy)C_{wf}T_{L1}x+C_{H1}E_{H1}x(T_{L1}\right.\\&\left.\times xy-T_{H3})]+x[C_{wf}T_{H1}(y-1)+C_{H1}E_{H1}(T_{H3}-T_{H1}y)]\right\}\end{aligned}}{\begin{aligned}&C_{wf}T_0xy-C_{wf}T_0x(1-E_H-E_L+E_HE_L)\\&\quad C_{wf}T_{H3}(1-E_H-E_L+E_HE_L)(E_{H1}-1)+C_{wf}\\&\quad\times y[E_HE_{H1}T_{H1}-T_{H3}(E_{H1}-1)]+E_{H1}E_LT_{L1}xy\\&\quad\times(1-E_H)\end{aligned}}
$$
$$
-\dfrac{C_{H1}}{C_{wf}T_0}\ln\dfrac{C_{wf}[T_{H3}y-T_{H3}(1-E_H)(1-E_L)]}{\vphantom{x}}
$$
$$
-\dfrac{C_H}{C_{wf}T_0}\ln\!\left\{1+\dfrac{E_HC_{wf}(T_{H1}-E_L\times T_{H1}-T_{H1}y+E_LT_{L1}xy)}{T_{H1}C_H[y-(1-E_H)(1-E_L)]}\right\}
$$
$$
-\dfrac{C_L}{C_{wf}T_0}\ln\!\left\{1+\dfrac{E_LC_{wf}[E_H(T_{H1}-T_{L1}x)-T_{L1}x(y-1)]}{C_LT_{L1}[(E_H-1)(1-E_L)x+xy]}\right\}
\tag{47}
$$

4. When $E_{H1}=0$, Equations (30)–(34) can be simplified into the performance indicators of the simple irreversible BCY coupled to VTHRs [79], whose $T-s$ diagram is shown in Figure 3d:

$$
\bar{W} = \dfrac{\begin{aligned}&C_{Lmin}C_{wf}E_LT_{L1}x+C_{Hmin}E_HT_{H1}\left\{C_{Lmin}E_L[\eta_t(x-1)-x]+C_{wf}x\right\}\\&+a_3\left\{C_{Lmin}(C_{wf}-C_{Hmin}E_H)E_L[\eta_t(x-1)-x]-C_{Hmin}C_{wf}E_Hx\right\}\end{aligned}}{C_{wf}^2T_0x}
\tag{48}
$$

$$
\eta = \dfrac{\begin{aligned}&a_3\left\{C_{Hmin}C_{wf}E_Hx-C_{Lmin}E_L(C_{wf}-C_{Hmin}E_H)[\eta_t(x-1)-x]\right\}-C_{Lmin}\\&\times C_{wf}E_LT_{L1}x+C_{Hmin}E_HT_{H1}[C_{Lmin}E_L(\eta_t+x-\eta_tx)-C_{wf}x]\end{aligned}}{xC_{Hmin}E_HC_{wf}(a_3-T_{H1})}
\tag{49}
$$

$$
\bar{P} = \dfrac{\begin{aligned}&\left\{-a_3[\eta_t(x-1)-x](C_{wf}-C_{Hmin}E_H)(C_{wf}-C_{Lmin}E_L)-C_{Hmin}E_HT_{H1}[\eta_t(x-1)-x]\right.\\&\left.\times(C_{wf}-C_{Lmin}E_L)+C_{Lmin}C_{wf}E_LT_{L1}x\right\}\left\{a_3\left\{C_{Lmin}E_L[\eta_t(x-1)-x](C_{wf}-C_{Hmin}E_H)\right.\right.\\&\left.-C_{Hmin}C_{wf}E_Hx\right\}+C_{Hmin}E_HT_{H1}\left\{C_{Lmin}E_L[\eta_t(x-1)-x]+C_{wf}x\right\}\left.+C_{Lmin}C_{wf}E_LT_{L1}x\right\}\end{aligned}}{-C_{wf}^3T_0x[\eta_t(x-1)-x][a_3(C_{wf}-C_{Hmin}E_H)+C_{Hmin}E_HT_{H1}]}
\tag{50}
$$

$$
\bar{E} = \dfrac{\begin{aligned}&\left\{C_{Lmin}C_{wf}E_LT_{L1}x+C_{Hmin}E_HT_{H1}\left\{C_{Lmin}E_L[\eta_t(x-1)-x]+C_{wf}x\right\}+a_3\left\{C_{Lmin}(C_{wf}\right.\right.\\&\left.-C_{Hmin}E_H)E_L[\eta_t(x-1)-x]-C_{Hmin}C_{wf}E_Hx\right\}\right\}/(T_0x)-C_{wf}\left\{C_H\ln[1+C_{Hmin}E_H\right.\\&\times(a_3-T_{H1})/(C_HT_{H1})]+C_L\ln\left\{1+C_{Lmin}E_L\left\{a_3C_{wf}\eta_t+C_{Hmin}E_H(a_3-T_{H1})[\eta_t(x-1)\right.\right.\\&\left.\left.\left.-x]-C_{wf}[a_3(\eta_t-1)+T_{L1}]x\right\}/(C_LC_{wf}T_{L1}x)\right\}\right\}\end{aligned}}{C_{wf}^2}
\tag{51}
$$

where

$$
a_3 = \dfrac{(\eta_c+x-1)\left\{C_{Lmin}C_{wf}E_LT_{L1}x-C_{Hmin}E_HT_{H1}(C_{wf}-C_{Lmin}E_L)[(\eta_t-1)x-\eta_t]\right\}}{\begin{aligned}&C_{Hmin}C_{Lmin}E_HE_L(\eta_c+x-1)(\eta_tx-x-\eta_t)+C_{wf}^2[x-x^2+\eta_t(\eta_c+x-1)(x\\&-1)]-C_{wf}(\eta_c+x-1)(E_HC_{Hmin}+E_LC_{Lmin})\times[(\eta_t-1)x-\eta_t]\end{aligned}}
\tag{52}
$$

5. When $E_{H1}=0$ and $C_H=C_L\to\infty$, Equations (30)–(34) can be simplified into the performance indicators of the simple irreversible BCY coupled to CTHRs [76], whose $T-s$ diagram is shown in Figure 3e:

$$
\bar{W} = \dfrac{E_LT_{L1}x-a_4\{(E_H-1)E_L[\eta_t(x-1)-x]+E_Hx\}+E_HT_{H1}[E_L\eta_t(x-1)+x-E_Lx]}{T_0x}
\tag{53}
$$

$$\eta = \frac{\begin{array}{c} a_4(E_H-1)E_L[\eta_t(x-1)-x]+a_4E_Hx-E_HT_{H1}x-E_LT_{L1}x \\ +E_HE_LT_{H1}(\eta_t+x-\eta_tx) \end{array}}{xE_H(a_4-T_{H1})} \tag{54}$$

$$\overline{P} = \frac{\begin{array}{c} \{a_4(E_H-1)(E_L-1)[\eta_t(x-1)-x]-E_HT_{H1}(E_L-1)[\eta_t(x \\ -1)-x]-E_LT_{L1}x\}\{a_4(E_H-1)E_L[\eta_t(x-1)-x]+a_4E_Hx \\ -E_HT_{H1}x-E_LT_{L1}x+E_HE_LT_{H1}(\eta_t+x-\eta_tx)\} \end{array}}{T_0[a_4(E_H-1)-E_HT_{H1}][\eta_t(x-1)-x]x} \tag{55}$$

$$\overline{E} = \begin{array}{c} \{E_LT_{L1}x-a_4\{E_L(E_H-1)[\eta_t(x-1)-x]+E_Hx\}+E_HT_{H1}[E_L\eta_t(x \\ -1)+x-E_Lx]\}/(T_0x)-C_H\ln[1+C_{wf}E_H(a_4-T_{H1})/(C_HT_{H1})]/C_{wf} \\ -C_L\ln\left\{1+C_{wf}E_L\{a_4(E_H-1)[\eta_t(x-1)-x]-T_{L1}x+E_HT_{H1}(\eta_t+x \\ -\eta_tx)\}/(C_LT_{L1}x)\right\}/C_{wf} \end{array} \tag{56}$$

where

$$a_4 = \frac{(\eta_c+x-1)E_HT_{H1}(E_L-1)[\eta_t(x-1)-x]+E_LT_{L1}x\}}{(E_H-1)(E_L-1)(x-1)(\eta_c+x-1)\eta_t-x[x-1+E_H(E_L-1)(\eta_c+x-1)-E_L(\eta_c+x-1)]} \tag{57}$$

6. When $E_{H1}=0$ and $\eta_c=\eta_t=1$, Equations (30)–(34) can be simplified into the performance indicators of the simple endoreversible BCY coupled to VTHRs [78], whose $T-s$ diagram is shown in Figure 3f:

$$\overline{W} = \frac{C_{Hmin}C_{Lmin}E_HE_L(-1+x)(T_{H1}-T_{L1}x)}{T_0x[C_{Lmin}C_{wf}E_L+C_{Hmin}E_H(C_{wf}-C_{Lmin}E_L)]} \tag{58}$$

$$\eta = (x-1)/x \tag{59}$$

$$\overline{P} = \frac{\begin{array}{c} C_{Hmin}C_{Lmin}E_HE_L(-1+x)(T_{H1}-T_{L1}x)[C_{Hmin}E_H(C_{wf} \\ -C_{Lmin}E_L)T_{H1}+C_{Lmin}C_{wf}E_LT_{L1}x] \end{array}}{\begin{array}{c} T_0x[C_{Lmin}C_{wf}E_L+C_{Hmin}E_H(C_{wf}-C_{Lmin}E_L)][C_{Lmin}C_{wf} \\ \times E_LT_{L1}x+C_{Hmin}E_H(C_{wf}T_{H1}-C_{Lmin}E_LT_{L1}x)] \end{array}} \tag{60}$$

$$\overline{E} = \frac{\begin{array}{c} \frac{C_{Hmin}C_{Lmin}C_{wf}E_HE_L(x-1)(T_{H1}-T_{L1}x)}{[C_{Lmin}C_{wf}E_L+C_{Hmin}E_H(C_{wf}-C_{Lmin}EL)]T_0x}-C_H\ln[1+\frac{C_{Hmin}C_{Lmin}C_{wf}E_HE_L(T_{L1}x-T_{H1})}{C_H[C_{Lmin}C_{wf}E_L+C_{Hmin}E_H(C_{wf}-C_{Lmin}E_L)]T_{H1}}] \\ -C_L\ln\{\frac{C_LC_{Lmin}C_{wf}E_LT_{L1}x+C_{Hmin}E_H[C_LC_{wf}T_{L1}x+C_{Lmin}E_L(C_{wf}T_{H1}-C_LT_{L1}x-C_{wf}T_{L1}x)]}{C_L[C_{Lmin}C_{wf}E_L+C_{Hmin}E_H(C_{wf}-C_{Lmin}E_L)]T_{L1}x}\} \end{array}}{C_{wf}} \tag{61}$$

7. When $E_{H1}=0$, $\eta_c=\eta_t=1$ and $C_H=C_L\to\infty$, Equations (30)–(34) can be simplified into the performance indicators of the simple endoreversible BCY coupled to CTHRs [77], whose $T-s$ diagram is shown in Figure 3g:

$$\overline{W} = \frac{E_HE_L(-1+x)(T_{L1}x-T_{H1})}{[E_H(E_L-1)-E_L]T_0x} \tag{62}$$

$$\eta = (x-1)/x \tag{63}$$

$$\overline{P} = \frac{E_HE_L(x-1)(T_{L1}x-T_{H1})[E_H(E_L-1)T_{H1}-E_LT_{L1}x]}{T_0x(E_HE_LT_{L1}x-E_HT_{H1}-E_LT_{L1}x)[E_H(E_L-1)-E_L]} \tag{64}$$

$$\overline{E} = \frac{\begin{array}{c} C_{wf}E_HE_L(x-1)(T_{L1}x-T_{H1})+C_HT_0x(E_H+E_L-E_HE_L)\ln\left\{1-C_{wf}E_H \\ \times E_L(T_{H1}-T_{L1}x)/[C_HT_{H1}(E_H+E_L-E_HE_L)]\right\}+C_LT_0x(E_H+E_L-E_H \\ \times E_L)\ln[1+C_{wf}E_HE_L(T_{H1}-T_{L1}x)/(C_L(E_H+E_L-E_HE_L)T_{L1}x)] \end{array}}{C_{wf}[E_H(E_L-1)-E_L]T_0x} \tag{65}$$

8. When $E_H = E_L = 0$, $\eta_c = \eta_t = 1$ and $C_{wf} \to \infty$, the cycle in this paper can become the endoreversible Carnot cycle coupled to VTHRs [14], whose $T - s$ diagram is shown in Figure 3h. However, Equations (30), (33), and (34) need to be de-dimensionalized to simplify to W, P and E of the endoreversible Carnot cycle coupled to VTHRs. The performance indicators of the cycle are:

$$W = \frac{C_H C_L E_H E_L (x-1)(T_{H1} - T_{L1}x)}{x(C_H E_H + C_L E_L)} \tag{66}$$

$$\eta = (x-1)/x \tag{67}$$

$$P = \frac{C_H C_L E_H E_L (x-1)(T_{H1} - T_{L1}x)}{x(C_H E_H + C_L E_L)} \tag{68}$$

$$\begin{aligned} E = {} & \frac{C_H C_L E_H E_L (x-1)(T_{H1} - T_{L1}x)}{(C_H E_H + C_L E_L)x} - C_H T_0 \ln\left[1 + \frac{C_L E_H E_L (T_{L1}x - T_{H1})}{(C_H E_H + C_L E_L)T_{H1}}\right] \\ & - C_L T_0 \ln\left[\frac{C_H E_H E_L T_{H1} + C_H E_H T_{L1}x + C_L E_L T_{L1}x - C_H E_H E_L T_{L1}x}{C_H E_H T_{L1}x + C_L E_L T_{L1}x}\right] \end{aligned} \tag{69}$$

9. When $E_H = E_L = 0$, $\eta_c = \eta_t = 1$ and $C_{H1} = C_L = C_{wf} \to \infty$, the cycle in this paper can become the endoreversible Carnot cycle coupled to CTHRs [12], whose $T - s$ diagram is shown in Figure 3i. However, Equations (30), (33), and (34) also need to be de-dimensionalized to simplify to W, P and E of the cycle [12,74,82]. The performance indicators of the cycle are:

$$W = \frac{U_H U_L (-1 + x)(T_{H1} - T_{L1}x)}{(U_H + U_L)x} \tag{70}$$

$$\eta = (x-1)/x \tag{71}$$

$$P = \frac{U_H U_L (-1 + x)(T_{H1} - T_{L1}x)}{(U_H + U_L)x} \tag{72}$$

$$E = \frac{U_H U_L (T_{H1} - T_{L1}x)[(T_0 + T_{H1})T_{L1}x - T_{H1}(T_0 + T_{L1})]}{T_{H1} T_{L1}(U_H + U_L)x} \tag{73}$$

10. When $E_H = E_L = 0$, $\eta_c = \eta_t = 1$, $C_{H1} = C_L = C_{wf} \to \infty$, and $U_L \to \infty$, the cycle in this paper can become the endoreversible Novikov cycle coupled to CTHRs [11], whose $T - s$ diagram is shown in Figure 3j. However, Equations (30), (33), and (34) also need to be de-dimensionalized to simplify to W, P and E of the cycle [11]. The performance indicators of the cycle are:

$$W = \frac{U_H(x-1)(T_{H1} - T_{L1}x)}{x} \tag{74}$$

$$\eta = (x-1)/x \tag{75}$$

$$P = \frac{U_H(x-1)(T_{H1} - T_{L1}x)}{x} \tag{76}$$

$$E = \frac{U_H(T_{H1} - T_{L1}x)[T_{H1}T_{L1}(x-1) + T_0(T_{L1}x - T_{H1})]}{T_{H1}T_{L1}x} \tag{77}$$

11. Through comparison with the results in Refs [11–14,59,76–79,99], it is found that the results of this paper are consistent with those in Refs [11–14,59,76–79,99], which further illustrates the accuracy of the model established in this paper. In particular, when the powers in Equations (58), (62), (66), (70), and (74) take the maximum values, namely $x = \sqrt{T_{H1}/T_{L1}}$, the efficiencies at the maximum power point, Equations (59), (63), (67), (71), and (75) are $\eta = 1 - \sqrt{T_{L1}/T_{H1}}$, which was derived in Refs. [10–12] by Moutier [10], Novikov [11], and Curzon and Ahlborn [12]. One can see that the results of this paper include the Novikov–Curzon–Ahlborn efficiency.

12. FTT is the further extension of conventional irreversible thermodynamics. The cycle model established by Curzon and Ahlborn [12] was a reciprocating Carnot cycle, and the finite time was its major feature. The methods used for solving the FTT problem are usually variational principle and optimal control theory. Therefore, such problems of extremal of thermodynamic processes were first named as FTT by Andresen et al. [132] and as Optimization Thermodynamics or Optimal Control in Problems of Extremals of Irreversible Thermodynamic Processes by Orlov and Rudenko [133]. When the research object was extended from reciprocating devices characterized by finite-time to the steady state flow devices characterized by finite-size, one realizes that the physical property of the problems is the heat transfer owing to temperature deference. Therefore, Grazzini [14] termed it Finite Temperature Difference Thermodynamics, and Lu [134] termed it Finite Surface Thermodynamics. In fact, the works performed by Moutier [10] and Novikov [11] were also steady state flow device models. Bejan introduced the effect of temperature difference heat transfer on the total entropy generation of the systems, taking the entropy generation minimization as the optimization objective for designing thermodynamic processes and devices, termed "Entropy Generation Minimization" or "Thermodynamic Optimization" [15,135]. For the steady state flow device models, Feidt [136–146] termed it Finite Physical Dimensions Thermodynamics (FPDT). The model established herein is closer to FPDT. For both reciprocating model and steady state flow model, the suitable name may be thermodynamics of finite size devices and finite time processes, as Bejan termed it [15,135]. According to the idiomatic usage, the theory is termed FTT in this paper.

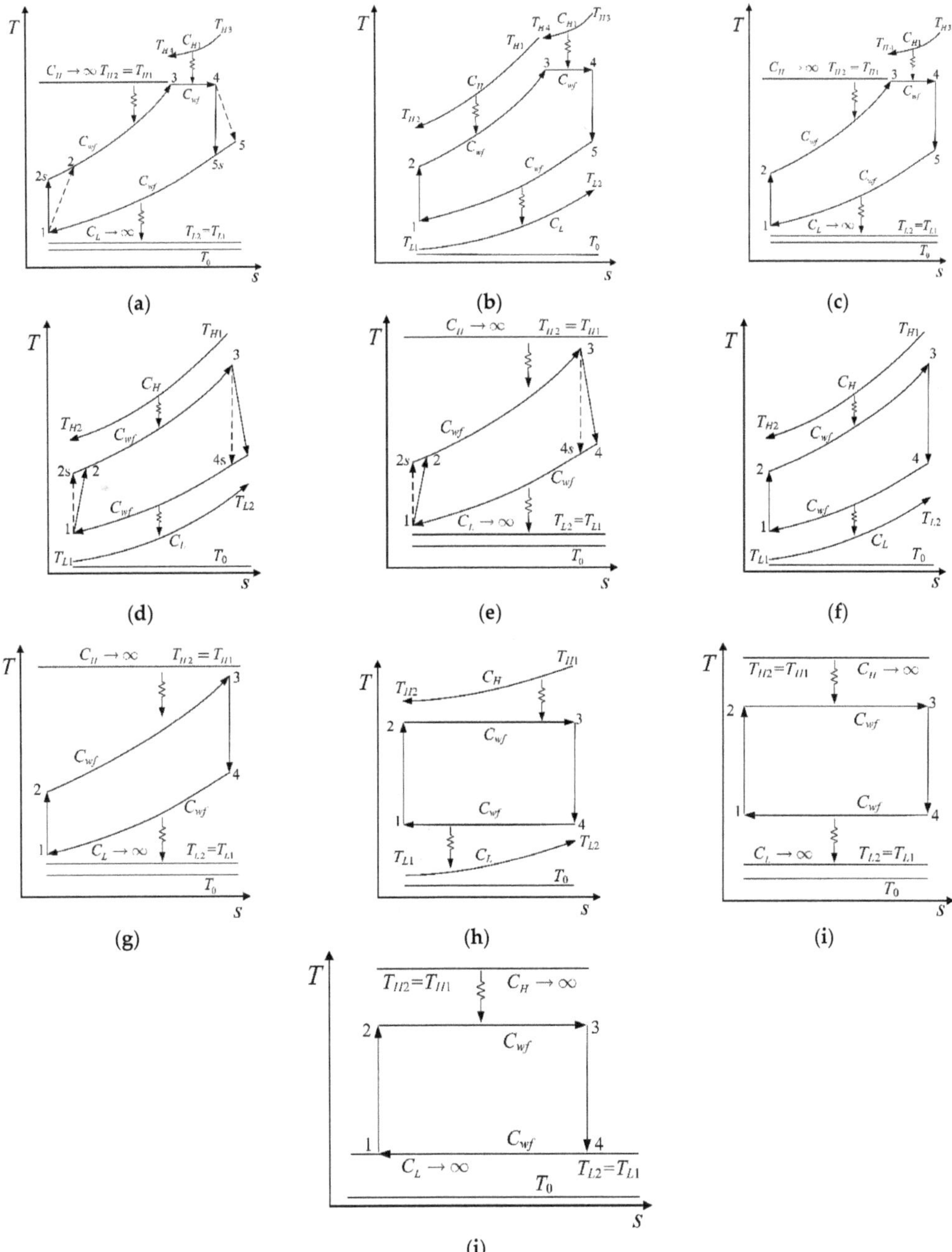

Figure 3. Diagrams of (**a**) irreversible simple BCY with an IHP and coupled to CTHRs; (**b**) endoreversible simple BCY with an IHP and coupled to VTHRs; (**c**) endoreversible simple BCY with an IHP and coupled to CTHRs; (**d**) simple irreversi-ble BCY coupled to VTHRs; (**e**) simple irreversible BCY coupled to CTHRs; (**f**) simple endoreversible BCY coupled to VTHRs; (**g**) simple endoreversible BCY coupled to CTHRs; (**h**) endoreversible Carnot cycle coupled to VTHRs; (**i**) endoreversible Carnot cycle coupled to CTHRs; (**j**) endoreversible Novikov cycle coupled to CTHRs.

3. Analyses and Optimizations with Each Single Objective

3.1. Analyses of Each Single Objective

The impacts of the irreversibility on cycle performance indicators ($\overline{W}$, η, $\overline{P}$ and $\overline{E}$) are analyzed below. In numerical calculations, it is set that $C_L = C_H = 1.2$ kW/K, $C_{wf} = 1$ kW/K, $T_0 = 300$ K, $C_{H1} = 0.6$ kW/K, $k = 1.4$, $R_g = 0.287$ kJ/(kg · K), $E_H = E_{H1} = E_L = 0.9$, $C_p = 1.005$ kW/K, $\tau_{H1} = 4.33$, $\tau_{H3} = 5$ and $\tau_L = 1$.

Figures 4–6 present the relationships of $\overline{W}$, η, $\overline{P}$, $\overline{E}$, π_t and v_5/v_1 versus π with different η_t. As shown in Figures 4 and 5, $\overline{W}$, η, $\overline{P}$ and $\overline{E}$ increase and then decrease as π increases. In the same situation, $\overline{W}$, $\overline{E}$, $\overline{P}$ and η reach the maximum value successively. When $\eta_t = 0.7$ and $\pi = 32.3$, $\overline{W} = \overline{P} = 0$. If π keeps going up, $\overline{W}$ and $\overline{P}$ are going to go negative. $\overline{W}$, η, $\overline{P}$ and $\overline{E}$ increase as η_t increases. As π increases, $\overline{W}$, η, $\overline{P}$ and $\overline{E}$ are affected more significantly by η_t. As shown in Figure 6, π_t goes up but v_5/v_1 goes down as π goes up. π_t and v_5/v_1 decrease as η_t rises. It illustrates that the degree of the IHP is improved and the device's volume is reduced as η_t increases.

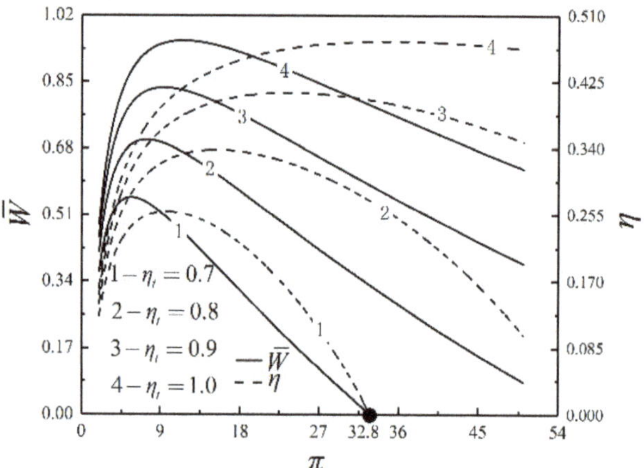

Figure 4. Relationships of $\overline{W}$ and η versus π with different η_t.

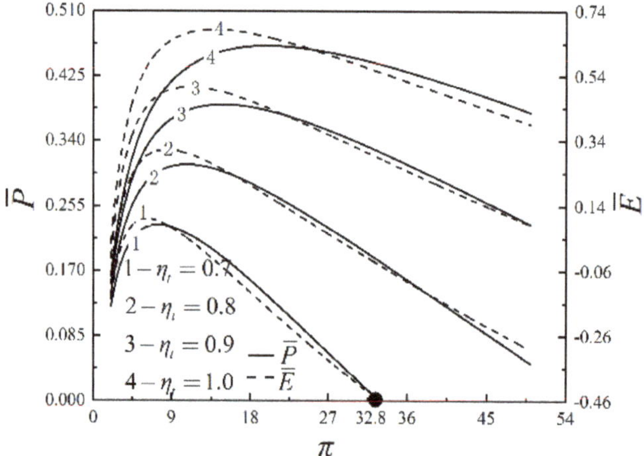

Figure 5. Relationships of $\overline{P}$ and $\overline{E}$ versus π with different η_t.

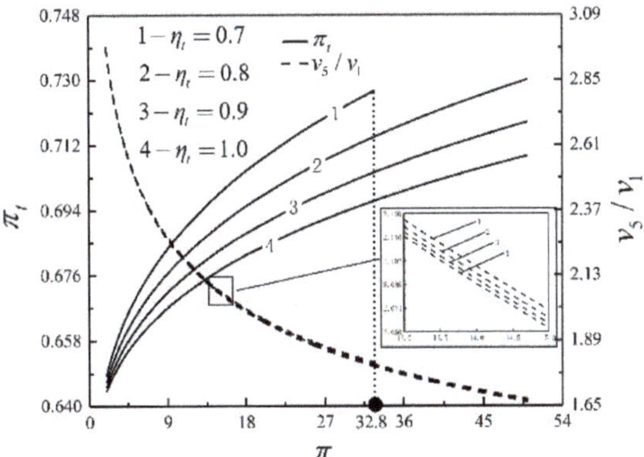

Figure 6. Relationships of π_t and v_5/v_1 versus π with different η_t.

By numerical calculations, the influences of η_c on $\overline{W}$, η, $\overline{P}$, $\overline{E}$ and π_t are the same as those of η_t on $\overline{W}$, η, $\overline{P}$, $\overline{E}$ and π_t. When $\eta_t = 0.7$ and $\pi = 32.8$, $\overline{W} = \overline{P} = 0$. However, the impacts of η_c on $\overline{W}$, η, $\overline{P}$ and $\overline{E}$ are less than those of η_t on $\overline{W}$, η, $\overline{P}$, $\overline{E}$. The effect of η_c on π_t is more significant than that of η_t on π_t. η_c has little effect on v_5/v_1. In the actual design process, it is suggested that η_t should be given priority.

To further explain the difference between the models in this paper and Ref. [101], the comparison of $\overline{W}$ under the variable and constant π is shown in Figure 7. As shown in Figure 7, $\overline{W}$ increases and then decreases as π increases in both cases; that is, the qualitative law is the same. However, there is an apparent quantitative difference between the two points. Under the constant π, $\overline{W}$ corresponding to the constant π is always greater than $\overline{W}$ conforming to the variable π. Similarly, there are quantitative differences in η, $\overline{P}$ and $\overline{E}$ under the variable and constant π. The model whose π is variable is more realistic.

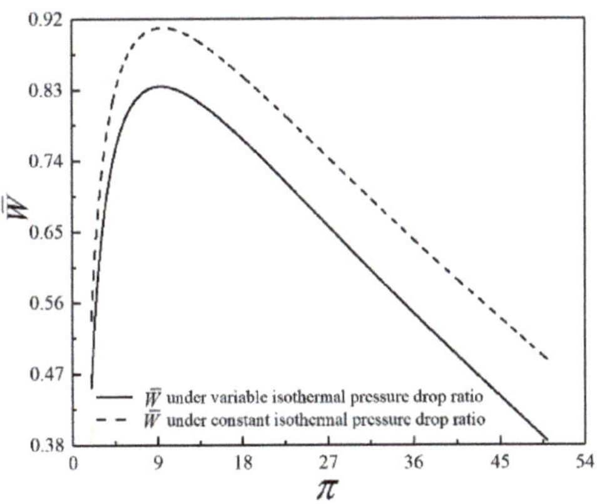

Figure 7. Comparison of $\overline{W}$ under the variable and constant π.

3.2. Performance Optimizations for Each Single Objective

With four performance indicators as the OPOs, respectively, the HCDs are optimized under the condition of given total heat conductance (U_T). The optimal results under different OPOs are compared. The HCDs among the RCC, CCC, and precooler are:

$$u_H = U_H/U_T, \quad u_{H1} = U_{H1}/U_T, \quad u_L = U_L/U_T \tag{78}$$

The HCDs are must larger than 0, the sum of them is 1, and $2 \leq \pi \leq 50$.

Figure 8 shows the flowchart of HCD optimization. The steps are as follows:

1. Enter the known data and the initial values of the HCDs.
2. The π_t is calculated according to Equation (13).
3. Judge whether the $\pi_t\pi$ and HCDs meet the constraints. If they are satisfied, perform step 4; if they are not satisfied, go back to step 1.
4. The performance indicator is solved.
5. Determine whether the inverse objective function is minimized by using the "fmincon" in MATLAB. If it is the smallest, perform step 6; if it is not the slightest, go back to step 1.
6. Calculate the other thermodynamic parameters, and the maximum of the performance indicator is obtained.

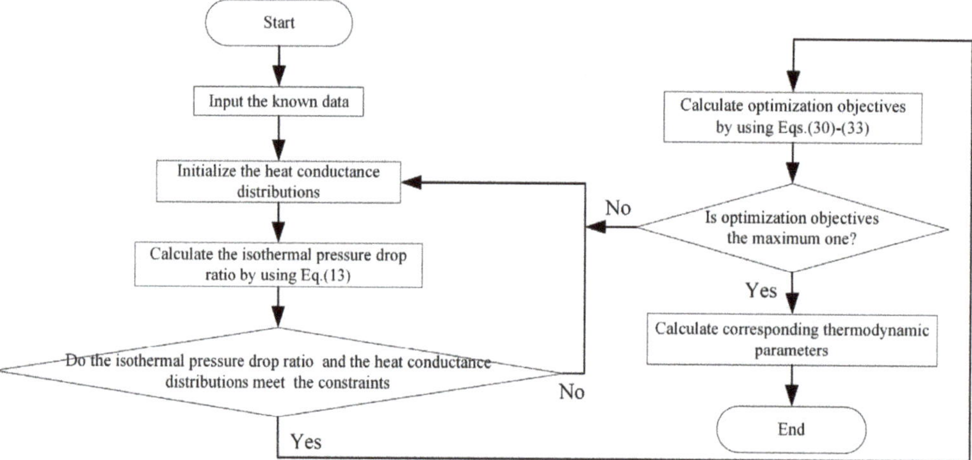

Figure 8. Flowchart of HCD optimization.

3.2.1. Optimizations of Each Single Objective

The optimization results of four performance indicators are similar. The optimization results with η as the performance indicator will be mainly discussed herein, while the results with $\overline{W}$, $\overline{P}$ and $\overline{E}$ as the performance indicators are briefly discussed. The relationships of the optimal thermal efficiency (η_{opt}) and the corresponding dimensionless power output ($\overline{W}_{\eta_{\text{opt}}}$) versus π are shown in Figure 9. The relationships of the corresponding dimensionless power density ($\overline{P}_{\eta_{\text{opt}}}$) and the corresponding dimensionless ecological function ($\overline{E}_{\eta_{\text{opt}}}$) versus π are demonstrated in Figure 10. As shown in Figures 9 and 10, $\overline{W}_{\eta_{\text{opt}}}$, η_{opt}, $\overline{P}_{\eta_{\text{opt}}}$ and $\overline{E}_{\eta_{\text{opt}}}$ first rise and then drops as π rises, which indicates a parabolic relationship with the downward opening. The corresponding isothermal pressure drop ratio ($(\pi_t)_{\eta_{\text{opt}}}$) and dimensionless maximum specific volume ($(v_5/v_1)_{\eta_{\text{opt}}}$) versus π are shown in Figure 11. $(\pi_t)_{\eta_{\text{opt}}}$ decreases and then increases as π increases. It indicates that there is a π_t that maximizes the degree of isothermal heating in the cycle. $(v_5/v_1)_{\eta_{\text{opt}}}$ decreases as π increases. The relationships of the HCDs ($(u_H)_{\eta_{\text{opt}}}$, $(u_{H1})_{\eta_{\text{opt}}}$ and $(u_L)_{\eta_{\text{opt}}}$) versus π are shown in Figure 12. As π increases, $(u_H)_{\eta_{\text{opt}}}$ decreases, $(u_{H1})_{\eta_{\text{opt}}}$ increases rapidly and then slowly, and $(u_L)_{\eta_{\text{opt}}}$ decreases first and then increases gradually.

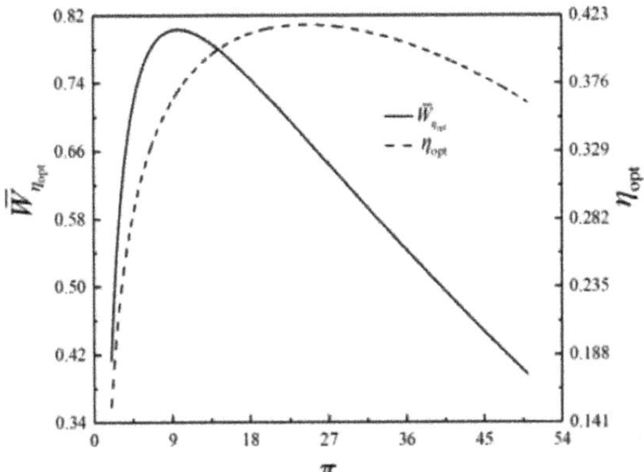

Figure 9. Relationships of $\overline{W}_{\eta_{\mathrm{opt}}}$ and η_{opt} versus π.

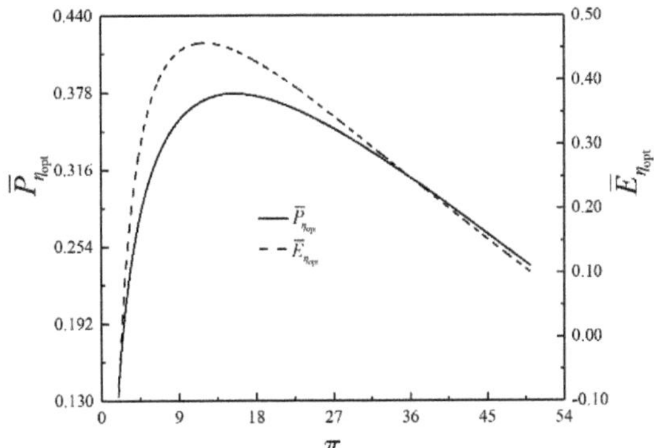

Figure 10. Relationships of $\overline{P}_{\eta_{\mathrm{opt}}}$ and $\overline{E}_{\eta_{\mathrm{opt}}}$ versus π.

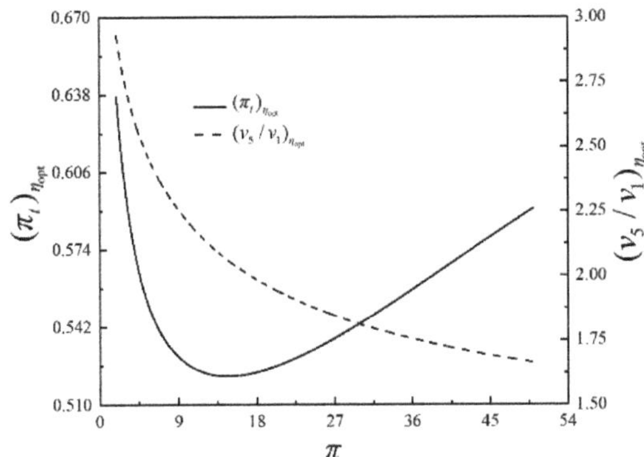

Figure 11. Relationships of $(\pi_t)_{\eta_{\mathrm{opt}}}$ and $(v_5/v_1)_{\eta_{\mathrm{opt}}}$ versus π.

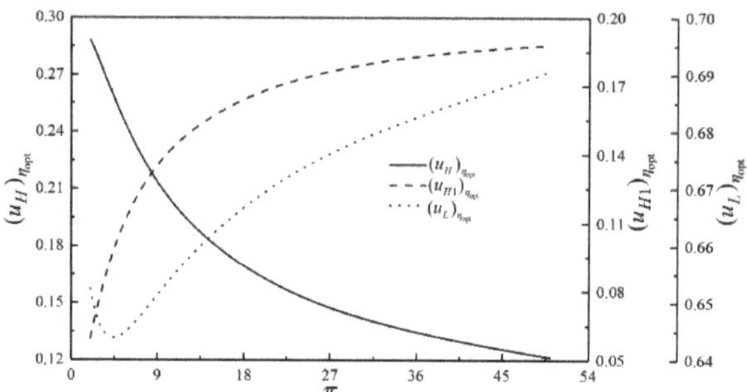

Figure 12. Relationships of $(u_H)_{\eta_{opt}}$, $(u_{H1})_{\eta_{opt}}$ and $(u_L)_{\eta_{opt}}$ versus π.

By numerical calculations, $\overline{W}_{opt}$, $\eta_{\overline{W}_{opt}}$, $\overline{P}_{\overline{W}_{opt}}$, $\overline{E}_{\overline{W}_{opt}}$, $\overline{W}_{\overline{P}_{opt}}$, $\eta_{\overline{P}_{opt}}$, $\overline{P}_{opt}$, $\overline{E}_{\overline{P}_{opt}}$, $\overline{W}_{\overline{E}_{opt}}$, $\eta_{\overline{E}_{opt}}$, $\overline{P}_{\overline{E}_{opt}}$ and $\overline{E}_{opt}$ increase first and then decrease as π increases. As π increases, $(\pi_t)_{\overline{W}_{opt}}$, $(\pi_t)_{\overline{P}_{opt}}$ and $(\pi_t)_{\overline{E}_{opt}}$ reduce first and then increase, and $(\pi_t)_{\overline{W}_{opt}}$, $(\pi_t)_{\overline{E}_{opt}}$, $(\pi_t)_{\eta_{opt}}$ and $(\pi_t)_{\overline{P}_{opt}}$ reached the minimum successively. As π increases, $(v_5/v_1)_{\overline{W}_{opt}}$, and $(v_5/v_1)_{\overline{E}_{opt}}$ decline, and their values have little difference. $(u_H)_{\overline{W}_{opt}}$, $(u_H)_{\eta_{opt}}$, $(u_H)_{\overline{P}_{opt}}$ and $(u_H)_{\overline{E}_{opt}}$ decrease as π increases, and $(u_H)_{\eta_{opt}}$ is always the smallest. $(u_{H1})_{\overline{W}_{opt}}$ and $(u_{H1})_{\overline{E}_{opt}}$ rise firstly and then tend to keep constant as π rises. $(u_{H1})_{\overline{P}_{opt}}$ first increases then decreases and finally tends to stay stable as π rises. $(u_L)_{\overline{W}_{opt}}$, $(u_L)_{\overline{P}_{opt}}$ and $(u_L)_{\overline{E}_{opt}}$ first increase rapidly and then slowly as π increases.

3.2.2. Influences of Temperature Ratios on Optimization Results

With η as the performance indicator, the influences of the temperature ratios on the optimization results are discussed. The relationship of the maximum thermal efficiency (η_{max}) versus τ_{H1} and τ_{H3} is shown in Figure 13. According to Figure 12, the surface is divided into three parts by line $\tau_{H3} = \tau_{H1} + 0.27$ (the correlation coefficient is $r_1 = 0.9969$) and $\tau_{H3} = 1.2\tau_{H1} + 0.1$ (the correlation coefficient is $r_2 = 1.0000$). τ_{H1} has little influence on η_{max}. When $\tau_{H3} < 1.2\tau_{H1} + 0.1$, η_{max} increases as τ_{H3} increases; when $\tau_{H3} > 1.2\tau_{H1} + 0.1$, τ_{H3} has little impact on η_{max}. It is recommended to magnify τ_{H1}.

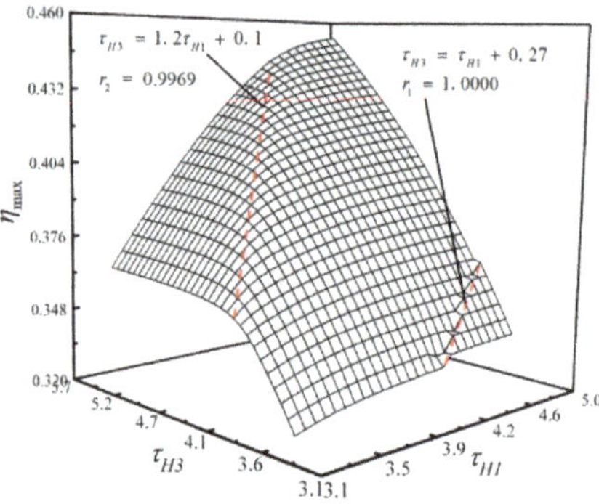

Figure 13. Relationships of η_{max} versus τ_{H1} and τ_{H3}.

By numerical calculations, the surface is divided into three parts by line $\tau_{H3} = 0.84\tau_{H1} + 0.41$ (the correlation coefficient is $r_1 = 0.9973$) and $\tau_{H3} = 1.2\tau_{H1} + 0.23$ (the correlation coefficient is $r_2 = 0.9988$) with $\overline{W}$ as the performance indicator. The surface is divided into three parts by line $\tau_{H3} = 0.78\tau_{H1} + 0.6$ (the correlation coefficient is $r_1 = 0.9574$) and $\tau_{H3} = 1.2\tau_{H1} + 0.33$ (the correlation coefficient is $r_2 = 0.9991$) with $\overline{P}$ as the performance indicator. The surface is divided into three parts by line $\tau_{H3} = 0.93\tau_{H1} + 0.058$ (the correlation coefficient is $r_1 = 0.9978$) and $\tau_{H3} = 1.1\tau_{H1} + 0.41$ (the correlation coefficient is $r_2 = 0.9990$) with $\overline{E}$ as the performance indicator. In practice, the difference between τ_{H1} and τ_{H3} should be controlled and should not be too large.

3.2.3. Influences of the Compressor and the Turbine's Irreversibilities on Optimization Results

With the four performance indicators as OPOs, respectively, the influences of η_c and η_t on optimization results are considered, and the thermodynamic parameters under various optimal performance indicators are compared. Figures 14 and 15 show relationships of $\overline{W}$ and π under various optimal performance indicators versus η_c and η_t, respectively $\overline{W}_{\max}$, $\overline{P}_{\max}$, and $\overline{E}_{\max}$ are the maximum dimensionless power output, maximum dimensionless power density, and maximum dimensionless ecological function, respectively. When $\overline{W}_{\max}$, $\eta_{\max}$, $\overline{P}_{\max}$, and $\overline{E}_{\max}$ are used as subscripts, they indicate the corresponding values at $\overline{W}_{\max}$, $\eta_{\max}$, $\overline{P}_{\max}$, and $\overline{E}_{\max}$ points.

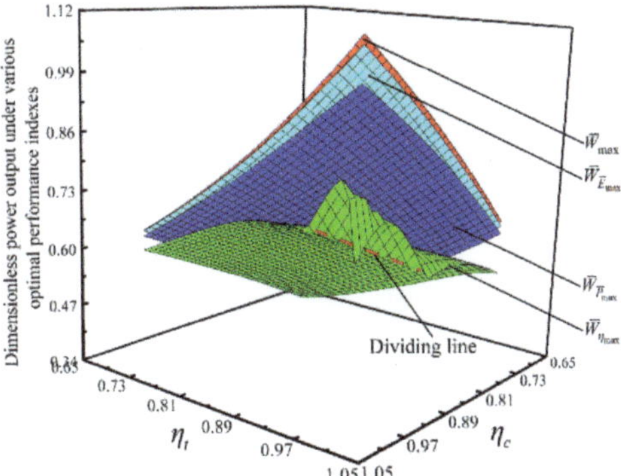

Figure 14. Relationships of $\overline{W}$ under various optimal performance indexes versus η_c and η_t.

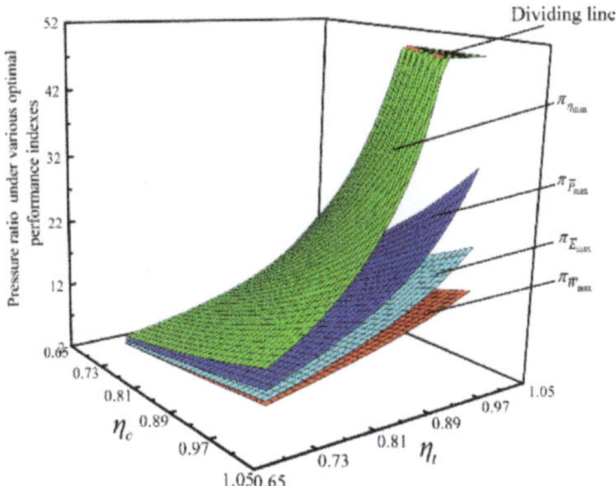

Figure 15. Relationships of π under various optimal performance indexes versus η_c and η_t.

As shown in Figure 14, $\overline{W}$ under various optimal performance indicators increases as η_c or η_t increases. When η_c and η_t both approach 1, $\overline{W}_{\eta_{\max}}$ first increases and then decreases as η_c or η_t increases. When $\eta_c = \eta_t = 1$, η rises monotonically as π gains, and there is no maximum value. In the case of the same η_c and η_t, there is $\overline{W}_{\max} > \overline{W}_{\overline{E}_{\max}} > \overline{W}_{\overline{P}_{\max}} > \overline{W}_{\eta_{\max}}$. As shown in Figure 15, π under various optimal performance indicators all increase as η_c or η_t increases. But the influence of η_t on π is more significant than that of η_c on π.

When η_c and η_t both approach 1, $\pi_{\eta_{max}}$ is always 50. Because the upper limit of π is 50. In the case of the same η_c and η_t, there is $\pi_{\eta_{max}} > \pi_{\overline{P}_{max}} > \pi_{\overline{E}_{max}} > \pi_{\overline{W}_{max}}$. The given range of π is $2 \leq \pi \leq 50$, so when $\pi = 50$, the trends of $\overline{W}_{\eta_{max}}$ and $\pi_{\eta_{max}}$ change significantly.

By numerical calculations, η, $\overline{P}$, and $\overline{E}$ under various optimal performance indicators increases as η_c or η_t increases. When η_c and η_t both approach 1, $\overline{P}_{\eta_{max}}$ and $\overline{E}_{\eta_{max}}$ first rises and then drops as η_c or η_t rises. In the same η_c and η_t, there are $\eta_{max} > \eta_{\overline{P}_{max}} > \eta_{\overline{E}_{max}} > \eta_{\overline{W}_{max}}$, $\overline{P}_{max} > \overline{P}_{\overline{E}_{max}} > \overline{P}_{\overline{W}_{max}} > \overline{P}_{\eta_{max}}$, (when η_c and η_t both tend to 1, the relationship does not work) and $\overline{E}_{max} > \overline{E}_{\overline{P}_{max}} > \overline{E}_{\overline{W}_{max}} > \overline{E}_{\eta_{max}}$ (the difference between $\overline{E}_{\overline{P}_{max}}$ and $\overline{E}_{\overline{W}_{max}}$ is very small).

The calculations also show that the thermal capacitance rate matchings among the VTHRs and working fluid have influences on the cycle performance. $\overline{W}_{max}$, η_{max}, $\overline{P}_{max}$, and $\overline{E}_{max}$ increase first and then keep constants as C_H/C_{wf} or C_{H1}/C_{wf} increases, and the effects of C_H/C_{wf} on $\overline{W}_{max}$, η_{max}, $\overline{P}_{max}$, and $\overline{E}_{max}$ are more significant than that of C_{H1}/C_{wf}.

4. Multi-Objective Optimization

4.1. Optimization Algorithm and Decision-Making Methods

It is impossible to achieve the maximums of $\overline{W}$, η, $\overline{P}$, and $\overline{E}$ under the same π. It shows that there is a contradiction among the four performance indicators. The multi-objective optimization problem is solved by applying the NSGA-II algorithm [99,100,102,105–125]. The detailed optimization process is shown in Figure 16. The Pareto frontier of the cycle performance is obtained by taking $\overline{W}$, η, $\overline{P}$, and $\overline{E}$ as OPOs, using the NSGA-II algorithm. The optimal scheme is selected by using the LINMAP, TOPSIS, and Shannon Entropy methods [99,102], and the algorithm of "gamultiobj" in MATLAB is based on the NSGA-II algorithm. The calculations are assisted by applying the "gamultiobj", and the corresponding Pareto frontier could be obtained. The parameter settings of "gamultiobj" are listed in Table 1.

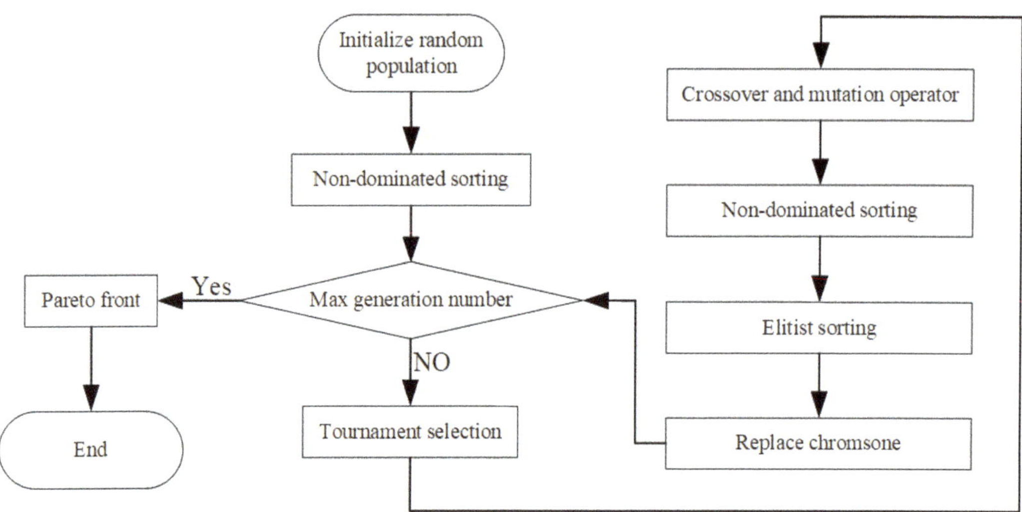

Figure 16. Flowchart of NSGA-II algorithm.

Table 1. Parameter settings of "gamultiobj".

Parameters	Values
Nvars	4
ParetoFraction	0.3
PopulationSize	300
Generations	500
CrossoverFraction	0.8

The positive and negative ideal points are the optimal and inferior schemes of each performance indicator. The LINMAP method is the Euclidian distance between each scheme and the positive ideal point, among which the one with the smallest distance is the best scheme. Suppose that the Pareto front contains n feasible solutions, and each viable solution contains m objective values $F_{ij}(1 \leq i \leq m$ and $1 \leq j \leq n)$. After normalizing F_{ij}, the value B_{ij} is:

$$B_{ij} = F_{ij} / \sqrt{\sum_{i=1}^{n} F_{ij}^2} \tag{79}$$

The weight of the j-th OPO is w_j^{LINMAP}, and the weighted value of B_{ij} is G_{ij}:

$$G_{ij} = w_j^{\text{LINMAP}} \cdot B_{ij} \tag{80}$$

The j-th objective of the positive ideal point is normalized and weighted, and the corresponding value is G_j^{positive}. The Euclidean distance between the i-th feasible solution and the positive ideal point is ED_i^+:

$$ED_i^+ = \sqrt{\sum_{j=1}^{m} \left(G_{ij} - G_j^{\text{positive}} \right)^2} \tag{81}$$

The best viable solution to the LINMAP method is i_{opt}:

$$i_{\text{opt}} \in \min\{ED_i^+\} \tag{82}$$

The TOPSIS method considers the Euclidean distance among each scheme and the positive and negative ideal points comprehensively, to further obtain the best scheme. The weight of the j-th OPO is w_j^{TOPSIS}, and the weighted value of B_{ij} is G_{ij}:

$$G_{ij} = w_j^{\text{TOPSIS}} \cdot B_{ij} \tag{83}$$

The j-th objective of the negative ideal point is normalized and weighted, and the corresponding value is G_j^{negative}. The Euclidean distance between the i-th feasible solution and the negative ideal point is ED_i^-:

$$ED_i^- = \sqrt{\sum_{j=1}^{m} \left(G_{ij} - G_j^{\text{negative}} \right)^2} \tag{84}$$

The best feasible solution of the TOPSIS method is i_{opt}:

$$i_{\text{opt}} \in \min\{ \frac{ED_i^-}{ED_i^+ + ED_i^-} \} \tag{85}$$

The Shannon Entropy method is a method to get the weight of multi-attribute decision-making.

After normalization of F_{ij}, P_{ij} is obtained:

$$P_{ij} = F_{ij} / \sum_{i=1}^{n} F_{ij} \tag{86}$$

The Shannon Entropy and weight of the j-th OPO are:

$$SE_j = -\frac{1}{\ln n} \sum_{i=1}^{n} P_{ij} \ln P_{ij} \tag{87}$$

$$w_j^{\text{Shannon Entropy}} = (1 - SE_j) / \sum_{j=1}^{n} (1 - SE_j) \tag{88}$$

The best feasible solution of the TOPSIS method is i_{opt}:

$$i_{\text{opt}} \in \min\left\{ P_{ij} \cdot w_j^{\text{Shannon Entropy}} \right\} \tag{89}$$

The deviation index D is defined as:

$$D = \frac{\sqrt{\sum_{j=1}^{m} \left(G_{i_{\text{opt}}j} - G_j^{\text{positive}} \right)^2}}{\sqrt{\sum_{j=1}^{m} \left(G_{i_{\text{opt}}j} - G_j^{\text{positive}} \right)^2} + \sqrt{\sum_{j=1}^{m} \left(G_{i_{\text{opt}}j} - G_j^{\text{positive}} \right)^2}} \tag{90}$$

In this paper, $w_j^{\text{LINMAP}} = w_j^{\text{TOPSIS}} = 1$ is chosen for the convenience of calculation.

4.2. Multi-Objective Optimization Results

Figure 17 shows the Pareto frontier and optimal schemes corresponding to the four objectives ($\overline{W}$, η, $\overline{P}$ and $\overline{E}$) optimization. The color on the Pareto frontier denotes the size of $\overline{E}$. To facilitate the observation of the changing relationships among the objectives, the pure red projection indicates the changing relationship between $\overline{W}$ and η. The pure green projection shows the changing relationship between $\overline{W}$ and $\overline{P}$, and the pure blue projection indicates the changing relationship between η and $\overline{P}$. It is easy to know that $\overline{W}$ and η, $\overline{W}$ and $\overline{P}$, η and $\overline{P}$ are all parabolic-like relationships with the opening downward. To analyze the influence of the corresponding optimization variables ($(u_H)_{\text{opt}}$, $(u_{H1})_{\text{opt}}$, $(u_L)_{\text{opt}}$ and π_{opt}) on cycle performance, the distributions of $(u_H)_{\text{opt}}$, $(u_{H1})_{\text{opt}}$, $(u_L)_{\text{opt}}$ and π_{opt} within the Pareto frontier's value range are shown in Figures 18–21. As shown in Figure 18, the value range of $(u_H)_{\text{opt}}$ is 0–1, but its distribution is between 0.167 and 0.272. As $(u_H)_{\text{opt}}$ increases, $\overline{W}$, $\overline{P}$, and $\overline{E}$ gradually increase, but η gradually decreases. As shown in Figure 19, the value range of $(u_{H1})_{\text{opt}}$ is 0–1, but its distribution is between 0.151 and 0.181. As $(u_{H1})_{\text{opt}}$ increases, $\overline{W}$, $\overline{P}$, and $\overline{E}$ gradually decrease, but the changing trend of η is not apparent. As shown in Figure 20, the value range of $(u_L)_{\text{opt}}$ is 0–1, but its distribution is between 0.568 and 0.662. As $(u_L)_{\text{opt}}$ increases, $\overline{W}$, $\overline{P}$, and $\overline{E}$ gradually decrease, but the changing trend of η is not apparent. As shown in Figure 21, the value range of π_{opt} is 2–50, but its distribution is between 9.692 and 24.426. As π_{opt} increases, $\overline{W}$ gradually decreases, η gradually increases, and $\overline{P}$ and $\overline{E}$ rise and then reduce.

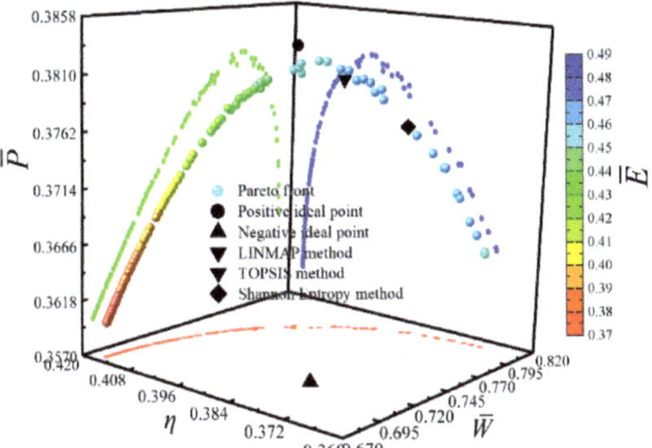

Figure 17. Pareto frontier and optimal schemes corresponding to the four objectives ($\overline{W}$, η, $\overline{P}$ and $\overline{E}$) optimization.

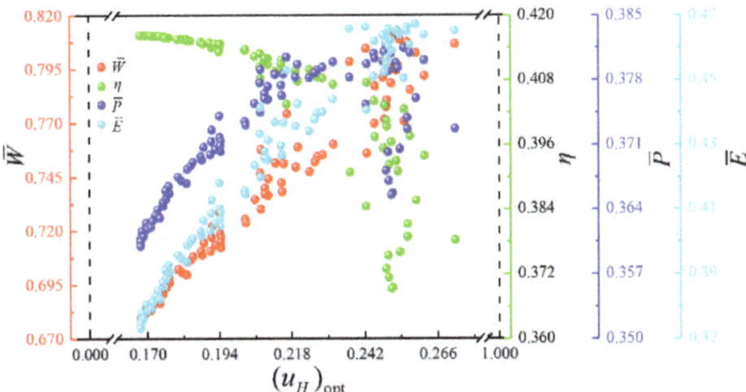

Figure 18. Distribution of $(u_H)_{\text{opt}}$ within the value range in the Pareto frontier.

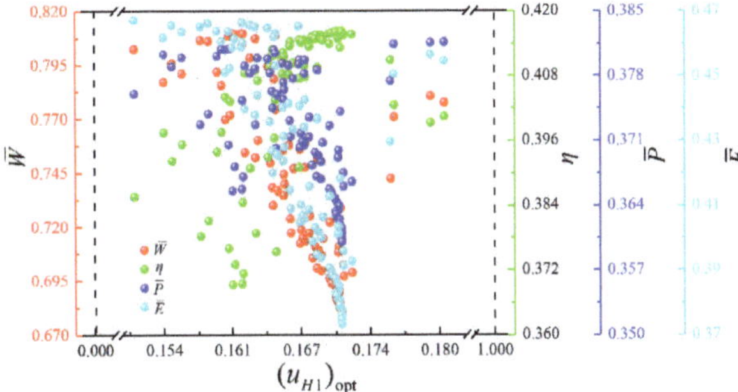

Figure 19. Distribution of $(u_{H1})_{\text{opt}}$ within the value range in the Pareto frontier.

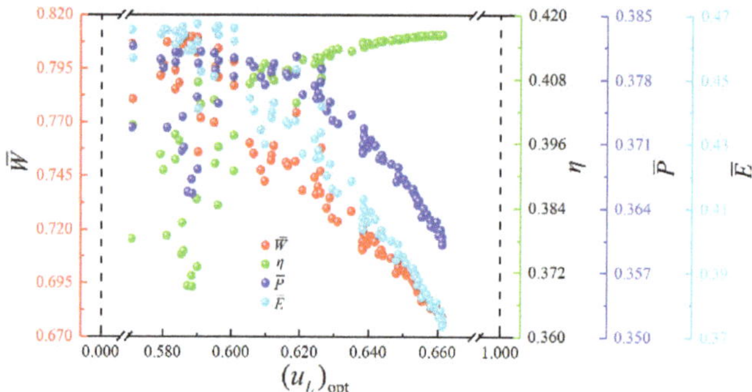

Figure 20. Distribution of $(u_L)_{\text{opt}}$ within the value range in the Pareto frontier.

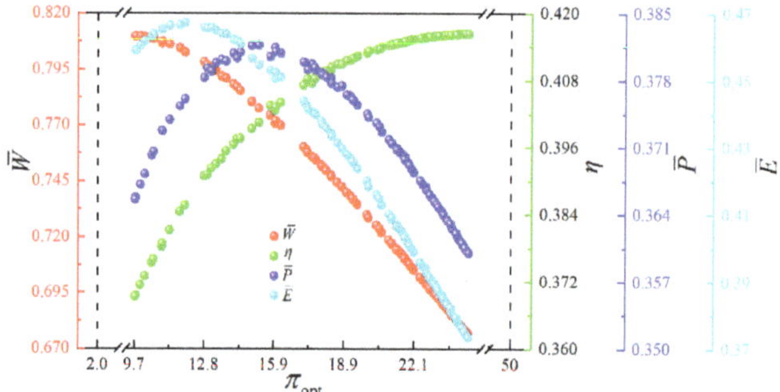

Figure 21. Distribution of π_{opt} within the value range in the Pareto frontier.

The Pareto frontier includes a series of non-inferior solutions, so the appropriate solution must be chosen according to the actual situation. The results of the triple- and double-objective optimizations are further discussed to compare the results of multi-objective optimizations more comprehensively. The comparison of the optimal schemes gotten by single- and double-, triple-, and quadruple-objective optimizations are listed in Table 2. The deviation index (D) is applied to represent the proximity between the optimal scheme and the positive ideal point. The appropriate optimal schemes are chosen by using the three methods. For the quadruple-objective optimization, $\overline{W}$, η, $\overline{P}$, and $\overline{E}$ corresponding to the positive ideal point are the maximum of the single-objective optimization. It indicates that the Pareto frontier includes all single-objective optimization results. The D obtained by the Shannon Entropy method is significantly smaller than that obtained by the LINMAP and TOPSIS methods. Simultaneously, it can be found that the D obtained by the Shannon Entropy method is the same as that with $\overline{E}$ as the OPO. For the triple-objective optimization, the triple-objective ($\overline{W}$, η and $\overline{E}$) optimization D obtained by the LINMAP or TOPSIS method is the smallest. For the double-objective optimization, the double-objective ($\overline{W}$ and $\overline{P}$) optimization D obtained by the LINMAP method is the smallest. For the single-objective optimization, the D corresponding to $\overline{E}_{\max}$ is the smallest. For single- and double-, triple-, and quadruple-objective optimizations, the double-objective ($\overline{W}$ and $\overline{P}$) optimization D obtained by the LINMAP method is the smallest.

Table 2. Comparison of the optimal schemes gotten by the single- and double-, triple-, and quadruple-objective optimizations.

OPOs	Decision Methods	Optimization Variables				Performance Indicators				Isothermal Pressure Drop Ratio	Deviation Indexes
		u_H	u_{H1}	u_L	π	$\overline{W}$	η	$\overline{P}$	$\overline{E}$	π_t	D
$\overline{W}, \eta, \overline{P}, \text{and } \overline{E}$	LINMAP	0.245	0.154	0.601	14.194	0.787	0.397	0.380	0.462	0.572	0.172
	TOPSIS	0.245	0.154	0.601	14.194	0.787	0.397	0.380	0.462	0.572	0.172
	Shannon Entropy	0.259	0.151	0.590	11.901	0.802	0.386	0.376	0.467	0.572	0.167
$\overline{W}, \eta, \text{and } \overline{P}$	LINMAP	0.230	0.167	0.603	14.261	0.787	0.398	0.381	0.461	0.557	0.170
	TOPSIS	0.231	0.167	0.602	14.115	0.788	0.397	0.380	0.462	0.557	0.168
	Shannon Entropy	0.246	0.167	0.587	12.008	0.803	0.386	0.377	0.466	0.559	0.165
$\overline{W}, \eta, \text{and } \overline{E}$	LINMAP	0.231	0.163	0.606	13.947	0.790	0.397	0.380	0.462	0.557	0.160
	TOPSIS	0.231	0.163	0.606	13.947	0.790	0.397	0.380	0.462	0.557	0.160
	Shannon Entropy	0.257	0.153	0.590	11.92	0.803	0.386	0.376	0.467	0.570	0.165
$\overline{W}, \overline{P} \text{ and } \overline{E}$	LINMAP	0.252	0.177	0.571	13.339	0.793	0.393	0.380	0.463	0.566	0.162
	TOPSIS	0.252	0.177	0.571	13.339	0.793	0.393	0.380	0.463	0.566	0.162
	Shannon Entropy	0.259	0.151	0.590	11.906	0.802	0.386	0.376	0.467	0.572	0.167
$\eta, \overline{P} \text{ and } \overline{E}$	LINMAP	0.241	0.170	0.589	17.016	0.761	0.406	0.380	0.444	0.575	0.319
	TOPSIS	0.241	0.170	0.589	17.016	0.761	0.406	0.380	0.444	0.575	0.319
	Shannon Entropy	0.245	0.169	0.585	16.674	0.764	0.405	0.381	0.447	0.577	0.297
$\overline{W} \text{ and } \eta$	LINMAP	0.230	0.169	0.601	14.293	0.787	0.398	0.381	0.461	0.557	0.170
	TOPSIS	0.230	0.169	0.601	14.293	0.787	0.398	0.381	0.461	0.557	0.170
	Shannon Entropy	0.248	0.168	0.585	12.061	0.802	0.387	0.377	0.466	0.560	0.162
$\overline{W} \text{ and } \overline{P}$	LINMAP	0.247	0.176	0.578	13.384	0.793	0.394	0.380	0.463	0.563	0.158
	TOPSIS	0.247	0.176	0.577	13.560	0.792	0.394	0.381	0.463	0.564	0.161
	Shannon Entropy	0.245	0.171	0.584	11.855	0.803	0.385	0.376	0.466	0.555	0.170
$\overline{W} \text{ and } \overline{E}$	LINMAP	0.258	0.154	0.589	11.765	0.803	0.385	0.376	0.467	0.570	0.169
	TOPSIS	0.258	0.154	0.589	11.765	0.803	0.385	0.376	0.467	0.570	0.169
	Shannon Entropy	0.259	0.152	0.589	11.902	0.802	0.386	0.376	0.467	0.572	0.167
$\eta \text{ and } \overline{P}$	LINMAP	0.232	0.192	0.576	16.452	0.765	0.405	0.381	0.446	0.562	0.295
	TOPSIS	0.235	0.193	0.572	16.156	0.768	0.404	0.381	0.447	0.563	0.279
	Shannon Entropy	0.241	0.196	0.563	15.603	0.772	0.402	0.381	0.450	0.564	0.255
$\eta \text{ and } \overline{E}$	LINMAP	0.237	0.1604	0.603	14.307	0.787	0.398	0.381	0.461	0.564	0.170
	TOPSIS	0.236	0.163	0.601	14.173	0.788	0.398	0.381	0.462	0.562	0.164
	Shannon Entropy	0.258	0.152	0.590	11.909	0.802	0.386	0.376	0.467	0.571	0.167
$\overline{P} \text{ and } \overline{E}$	LINMAP	0.257	0.166	0.578	13.483	0.792	0.394	0.380	0.464	0.572	0.160
	TOPSIS	0.257	0.165	0.578	13.386	0.793	0.393	0.380	0.464	0.572	0.161
	Shannon Entropy	0.258	0.154	0.588	12.054	0.802	0.387	0.377	0.467	0.571	0.161

Table 2. *Cont.*

OPOs	Decision Methods	Optimization Variables				Performance Indicators				Isothermal Pressure Drop Ratio	Deviation Indexes
		u_H	u_{H1}	u_L	π	$\overline{W}$	η	$\overline{P}$	$\overline{E}$	π_t	D
	$\overline{W}$	0.249	0.162	0.589	9.678	0.810	0.369	0.365	0.459	0.550	0.242
	η	0.152	0.174	0.674	24.542	0.672	0.416	0.358	0.369	0.532	0.783
	$\overline{P}$	0.251	0.183	0.567	15.149	0.777	0.400	0.382	0.454	0.571	0.225
	$\overline{E}$	0.259	0.151	0.590	11.903	0.802	0.386	0.376	0.467	0.572	0.167
Positive ideal point		——	——	——	——	0.810	0.416	0.382	0.467	0.810	——
Negative ideal point		——	——	——	——	0.677	0.369	0.360	0.373	0.677	——

5. Conclusions

Based on FTT, an improved irreversible closed modified simple BCY model with one IHP and coupled to VTHRs is established and optimized with four performance indicators as OPOs, respectively. The optimization results are compared, and the influences of compressor and turbine efficiencies on optimization results are analyzed. Finally, the cycle is optimized, and the corresponding Pareto frontier is gained by adopting the NSGA-II algorithm. Based on three different methods, the optimal scheme is gotten from the Pareto frontier. The results obtained in this paper reveal the original results in Refs. [10–12], which were the initial work of the FTT theory. The main results are summarized:

1. For the single-objective analyses and optimizations, performance indicators all rise as η_c and η_t rise. The influences of η_t on four performance indicators are greater than those of η_c. $\overline{W}$ of the models in this paper increase and then decrease as π increases in both cases; that is, the qualitative law is the same. However, there is an apparent quantitative difference between the two points. In practice, the difference between τ_{H1} and τ_{H3} should be controlled and not be too large. $\overline{P}$ and $\overline{E}$ are the trade-offs between $\overline{W}$ and η.

2. For single- and double-, triple-, and quadruple-objective optimizations, the Pareto frontier includes a series of non-inferior solutions. The appropriate solution could be chosen according to the actual situation. By comparison, it is found that the double-objective ($\overline{W}$ and $\overline{P}$) optimization D obtained by the LINMAP method is the smallest.

3. The optimization results gained in this paper could offer theoretical guidelines for the optimal designs of the gas turbine plants. In the next step, the improved closed intercooling regenerated modified BCY model with one IHP will be optimized with real gas as the working fluid, and the internal friction-based pressure drops during heating and cooling processes and other processes, as well as the heat leakage losses between the heat source and the environment, will be taken into account.

Author Contributions: Conceptualization: L.C. and H.F.; funding acquisition: L.C.; methodology: C.T.; software: C.T. and Y.G.; validation: L.C. and Y.G; writing—original draft: C.T. and H.F.; writing—review and editing: L.C. All authors have read and agreed to the published version of the manuscript.

Funding: This work is supported by the National Natural Science Foundation of China (Grant No. 51779262).

Acknowledgments: The authors wish to thank the reviewers for their careful, unbiased, and constructive suggestions, which led to this revised manuscript.

Nomenclature

a, x, y	Intermediate variables
C	Thermal capacity rate (kW/K)
C_p	Specific heat at constant pressure (kJ/(kg·K))
E	Effectiveness of heat exchanger or ecological function (kW)
$\overline{E}$	Dimensionless ecological function
k	Specific heat ratio
M	Mach number
N	Number of the heat transfer unit
$\dot{Q}$	Heat absorbing rate or heat releasing rate (kW)
$\overline{P}$	Dimensionless power density
T	Temperature (K)
U	Heat conductance (kW/K)
u	Heat conductance distribution

$\overline{W}$	Dimensionless power output
Greek symbols	
η	Efficiency
π	Pressure ratio
τ	Temperature ratio
Subscripts	
H	Hot-side heat exchanger
L	Cold-side heat exchanger
wf	Working fluid
$1, 2, 3, 4, 5, 2s, 5s$	State points

Abbreviations

Brayton cycle	BCY
CCC	Convergent combustion chamber
CTHR	Constant-temperature heat reservoir
FPDT	Finite Physical Dimensions Thermodynamics
FTT	Finite time thermodynamics
HCD	Heat conductance distribution
IHP	Isothermal heating process
OPO	Optimization objective
RCC	Regular combustion chamber
VTHR	Variable-temperature heat reservoir

References

1. Wood, W.A. On the role of the harmonic mean isentropic exponent in the analysis of the closed-cycle gas turbine. *Proc. Inst. Mech. Eng. Part. A J. Power Energy* **1991**, *205*, 287–291. [CrossRef]
2. Cheng, K.L.; Qin, J.; Sun, H.C.; Li, H.; He, S.; Zhang, S.L.; Bao, W. Power optimization and comparison between simple recuperated and recompressing supercritical carbon dioxide Closed-Brayton-Cycle with finite cold source on hypersonic vehicles. *Energy* **2019**, *181*, 1189–1201. [CrossRef]
3. Hu, H.M.; Jiang, Y.Y.; Guo, C.H.; Liang, S.Q. Thermodynamic and exergy analysis of a S-CO2 Brayton cycle with various of cooling modes. *Energy Convers. Manag.* **2020**, *220*, 113110. [CrossRef]
4. Liu, H.Q.; Chi, Z.R.; Zang, S.S. Optimization of a closed Brayton cycle for space power systems. *Appl. Therm. Eng.* **2020**, *179*, 115611. [CrossRef]
5. Vecchiarelli, J.; Kawall, J.G.; Wallace, J.S. Analysis of a concept for increasing the efficiency of a Brayton cycle via isothermal heat addition. *Int. J. Energy Res.* **1997**, *21*, 113–127. [CrossRef]
6. Göktun, S.; Yavuz, H. Thermal efficiency of a regenerative Brayton cycle with isothermal heat addition. *Energy Convers. Manag.* **1999**, *40*, 1259–1266. [CrossRef]
7. Erbay, L.B.; Göktun, S.; Yavuz, H. Optimal design of the regenerative gas turbine engine with isothermal heat addition. *Appl. Energy* **2001**, *68*, 249–264. [CrossRef]
8. Jubeh, N.M. Exergy analysis and second law efficiency of a regenerative Brayton cycle with isothermal heat addition. *Entropy* **2005**, *7*, 172–187. [CrossRef]
9. El-Maksound, R.M.A. Binary Brayton cycle with two isothermal processes. *Energy Convers. Manag.* **2013**, *73*, 303–308. [CrossRef]
10. Moutier, J. *Éléments de Thermodynamique*; Gautier-Villars: Paris, France, 1872.
11. Novikov, I.I. The efficiency of atomic power stations (A review). *J. Nucl. Energy* **1957**, *7*, 125–128. [CrossRef]
12. Curzon, F.L.; Ahlborn, B. Efficiency of a Carnot engine at maximum power output. *Am. J. Phys.* **1975**, *43*, 22–24. [CrossRef]
13. Andresen, B. *Finite-Time Thermodynamics*; Physics Laboratory II; University of Copenhagen: Copenhagen, Denmark, 1983.
14. Grazzini, G. Work from irreversible heat engines. *Energy* **1991**, *16*, 747–755. [CrossRef]
15. Bejan, A. *Entropy Generation Minimization*; CRC Press: Boca Raton, FL, USA, 1996.
16. Chen, L.G.; Wu, C.; Sun, F.R. Finite time thermodynamic optimization or entropy generation minimization of energy systems. *J. Non-Equilib. Thermodyn.* **1999**, *24*, 327–359. [CrossRef]
17. Andresen, B. Current trends in finite-time thermodynamics. *Angew. Chem. Int. Ed.* **2011**, *50*, 2690–2704. [CrossRef] [PubMed]
18. Shittu, S.; Li, G.Q.; Zhao, X.D.; Ma, X.L. Review of thermoelectric geometry and structure optimization for performance enhancement. *Appl. Energy* **2020**, *268*, 115075. [CrossRef]
19. Berry, R.S.; Salamon, P.; Andresen, B. How it all began. *Entropy* **2020**, *22*, 908. [CrossRef] [PubMed]
20. Hoffman, K.H.; Burzler, J.; Fischer, A.; Schaller, M.; Schubert, S. Optimal process paths for endoreversible systems. *J. Non-Equilib. Thermodyn.* **2003**, *28*, 233–268. [CrossRef]

21. Zaeva, M.A.; Tsirlin, A.M.; Didina, O.V. Finite time thermodynamics: Realizability domain of heat to work converters. *J. Non-Equilib. Thermodyn.* **2019**, *44*, 181–191. [CrossRef]
22. Masser, R.; Hoffmann, K.H. Endoreversible modeling of a hydraulic recuperation system. *Entropy* **2020**, *22*, 383. [CrossRef]
23. Kushner, A.; Lychagin, V.; Roop, M. Optimal thermodynamic processes for gases. *Entropy* **2020**, *22*, 448. [CrossRef]
24. De Vos, A. Endoreversible models for the thermodynamics of computing. *Entropy* **2020**, *22*, 660. [CrossRef] [PubMed]
25. Masser, R.; Khodja, A.; Scheunert, M.; Schwalbe, K.; Fischer, A.; Paul, R.; Hoffmann, K.H. Optimized piston motion for an alpha-type Stirling engine. *Entropy* **2020**, *22*, 700. [CrossRef] [PubMed]
26. Chen, L.G.; Ma, K.; Ge, Y.L.; Feng, H.J. Re-optimization of expansion work of a heated working fluid with generalized radiative heat transfer law. *Entropy* **2020**, *22*, 720. [CrossRef] [PubMed]
27. Tsirlin, A.; Gagarina, L. Finite-time thermodynamics in economics. *Entropy* **2020**, *22*, 891. [CrossRef] [PubMed]
28. Tsirlin, A.; Sukin, I. Averaged optimization and finite-time thermodynamics. *Entropy* **2020**, *22*, 912. [CrossRef]
29. Muschik, W.; Hoffmann, K.H. Modeling, simulation, and reconstruction of 2-reservoir heat-to-power processes in finite-time thermodynamics. *Entropy* **2020**, *22*, 997. [CrossRef] [PubMed]
30. Insinga, A.R. The quantum friction and optimal finite-time performance of the quantum Otto cycle. *Entropy* **2020**, *22*, 1060. [CrossRef]
31. Schön, J.C. Optimal control of hydrogen atom-like systems as thermodynamic engines in finite time. *Entropy* **2020**, *22*, 1066. [CrossRef]
32. Andresen, B.; Essex, C. Thermodynamics at very long time and space scales. *Entropy* **2020**, *22*, 1090. [CrossRef]
33. Chen, L.G.; Ma, K.; Feng, H.J.; Ge, Y.L. Optimal configuration of a gas expansion process in a piston-type cylinder with generalized convective heat transfer law. *Energies* **2020**, *13*, 3229. [CrossRef]
34. Scheunert, M.; Masser, R.; Khodja, A.; Paul, R.; Schwalbe, K.; Fischer, A.; Hoffmann, K.H. Power-optimized sinusoidal piston motion and its performance gain for an Alpha-type Stirling engine with limited regeneration. *Energies* **2020**, *13*, 4564. [CrossRef]
35. Boikov, S.Y.; Andresen, B.; Akhremenkov, A.A.; Tsirlin, A.M. Evaluation of irreversibility and optimal organization of an integrated multi-stream heat exchange system. *J. Non-Equilib. Thermodyn.* **2020**, *45*, 155–171. [CrossRef]
36. Chen, L.G.; Feng, H.J.; Ge, Y.L. Maximum energy output chemical pump configuration with an infinite-low- and a finite-high-chemical potential mass reservoirs. *Energy Convers. Manag.* **2020**, *223*, 113261. [CrossRef]
37. Hoffmann, K.H.; Burzler, J.M.; Schubert, S. Endoreversible thermodynamics. *J. Non-Equilib. Thermodyn.* **1997**, *22*, 311–355.
38. Wagner, K.; Hoffmann, K.H. Endoreversible modeling of a PEM fuel cell. *J. Non-Equilib. Thermodyn.* **2015**, *40*, 283–294. [CrossRef]
39. Muschik, W. Concepts of phenominological irreversible quantum thermodynamics I: Closed undecomposed Schottky systems in semi-classical description. *J. Non-Equilib. Thermodyn.* **2019**, *44*, 1–13. [CrossRef]
40. Ponmurugan, M. Attainability of maximum work and the reversible efficiency of minimally nonlinear irreversible heat engines. *J. Non-Equilib. Thermodyn.* **2019**, *44*, 143–153. [CrossRef]
41. Raman, R.; Kumar, N. Performance analysis of Diesel cycle under efficient power density condition with variable specific heat of working fluid. *J. Non-Equilib. Thermodyn.* **2019**, *44*, 405–416. [CrossRef]
42. Schwalbe, K.; Hoffmann, K.H. Stochastic Novikov engine with Fourier heat transport. *J. Non-Equilib. Thermodyn.* **2019**, *44*, 417–424. [CrossRef]
43. Morisaki, T.; Ikegami, Y. Maximum power of a multistage Rankine cycle in low-grade thermal energy conversion. *Appl. Thermal Eng.* **2014**, *69*, 78–85. [CrossRef]
44. Yasunaga, T.; Ikegami, Y. Application of finite time thermodynamics for evaluation method of heat engines. *Energy Proc.* **2017**, *129*, 995–1001. [CrossRef]
45. Yasunaga, T.; Fontaine, K.; Morisaki, T.; Ikegami, Y. Performance evaluation of heat exchangers for application to ocean thermal energy conversion system. *Ocean Thermal Energy Convers.* **2017**, *22*, 65–75.
46. Yasunaga, T.; Koyama, N.; Noguchi, T.; Morisaki, T.; Ikegami, Y. Thermodynamical optimum heat source mean velocity in heat exchangers on OTEC. In Proceedings of the Grand Renewable Energy 2018, Yokohama, Japan, 17–22 June 2018.
47. Yasunaga, T.; Noguchi, T.; Morisaki, T.; Ikegami, Y. Basic heat exchanger performance evaluation method on OTEC. *J. Mar. Sci. Eng.* **2018**, *6*, 32. [CrossRef]
48. Fontaine, K.; Yasunaga, T.; Ikegami, Y. OTEC maximum net power output using Carnot cycle and application to simplify heat exchanger selection. *Entropy* **2019**, *21*, 1143. [CrossRef]
49. Yasunaga, T.; Ikegami, Y. Finite-time thermodynamic model for evaluating heat engines in ocean thermal energy conversion. *Entropy* **2020**, *22*, 211. [CrossRef]
50. Shittu, S.; Li, G.Q.; Zhao, X.D.; Ma, X.L.; Akhlaghi, Y.G.; Fan, Y. Comprehensive study and optimization of concentrated photovoltaic-thermoelectric considering all contact resistances. *Energy Convers. Manag.* **2020**, *205*, 112422. [CrossRef]
51. Feidt, M. Carnot cycle and heat engine: Fundamentals and applications. *Entropy* **2020**, *22*, 348. [CrossRef]
52. Feidt, M.; Costea, M. Effect of machine entropy production on the optimal performance of a refrigerator. *Entropy* **2020**, *22*, 913. [CrossRef] [PubMed]
53. Ma, Y.H. Effect of finite-size heat source's heat capacity on the efficiency of heat engine. *Entropy* **2020**, *22*, 1002. [CrossRef] [PubMed]
54. Rogolino, P.; Cimmelli, V.A. Thermoelectric efficiency of Silicon–Germanium alloys in finite-time thermodynamics. *Entropy* **2020**, *22*, 1116. [CrossRef] [PubMed]

55. Dann, R.; Kosloff, R.; Salamon, P. Quantum finite time thermodynamics: Insight from a single qubit engine. *Entropy* **2020**, *22*, 1255. [CrossRef] [PubMed]

56. Liu, X.W.; Chen, L.G.; Ge, Y.L.; Feng, H.J.; Wu, F.; Lorenzini, G. Exergy-based ecological optimization of an irreversible quantum Carnot heat pump with spin-1/2 systems. *J. Non-Equilib. Thermodyn.* **2021**, *46*, 61–76. [CrossRef]

57. Guo, H.; Xu, Y.J.; Zhang, X.J.; Zhu, Y.L.; Chen, H.S. Finite-time thermodynamics modeling and analysis on compressed air energy storage systems with thermal storage. *Renew. Sustain. Energy Rev.* **2021**, *138*, 110656. [CrossRef]

58. Smith, Z.; Pal, P.S.; Deffner, S. Endoreversible Otto engines at maximal power. *J. Non-Equilib. Thermodyn.* **2020**, *45*, 305–310. [CrossRef]

59. Chen, L.G.; Shen, J.F.; Ge, Y.L.; Wu, Z.X.; Wang, W.H.; Zhu, F.L.; Feng, H.J. Power and efficiency optimization of open Maisotsenko-Brayton cycle and performance comparison with traditional open regenerated Brayton cycle. *Energy Convers. Manag.* **2020**, *217*, 113001. [CrossRef]

60. Liu, H.T.; Zhai, R.R.; Patchigolla, K.; Turner, P.; Yang, Y.P. Analysis of integration method in multi-heat-source power generation systems based on finite-time thermodynamics. *Energy Convers. Manag.* **2020**, *220*, 113069. [CrossRef]

61. Feng, H.J.; Qin, W.X.; Chen, L.G.; Cai, C.G.; Ge, Y.L.; Xia, S.J. Power output, thermal efficiency and exergy-based ecological performance optimizations of an irreversible KCS-34 coupled to variable temperature heat reservoirs. *Energy Convers. Manag.* **2020**, *205*, 112424. [CrossRef]

62. Feng, J.S.; Gao, G.T.; Dabwan, Y.N.; Pei, G.; Dong, H. Thermal performance evaluation of subcritical organic Rankine cycle for waste heat recovery from sinter annular cooler. *J. Iron. Steel Res. Int.* **2020**, *27*, 248–258. [CrossRef]

63. Wu, Z.X.; Feng, H.J.; Chen, L.G.; Tang, W.; Shi, J.C.; Ge, Y.L. Constructal thermodynamic optimization for ocean thermal energy conversion system with dual-pressure organic Rankine cycle. *Energy Convers. Manag.* **2020**, *210*, 112727. [CrossRef]

64. Qiu, S.S.; Ding, Z.M.; Chen, L.G. Performance evaluation and parametric optimum design of irreversible thermionic generators based on van der Waals heterostructures. *Energy Convers. Manag.* **2020**, *225*, 113360. [CrossRef]

65. Miller, H.J.D.; Mehboudi, M. Geometry of work fluctuations versus efficiency in microscopic thermal machines. *Phys. Rev. Lett.* **2020**, *125*, 260602. [CrossRef]

66. Gonzalez-Ayala, J.; Roco, J.M.M.; Medina, A.; Calvo Hernández, A. Optimization, stability, and entropy in endoreversible heat engines. *Entropy* **2020**, *22*, 1323. [CrossRef]

67. Kong, R.; Chen, L.G.; Xia, S.J.; Li, P.L.; Ge, Y.L. Minimizing entropy generation rate in hydrogen iodide decomposition reactor heated by high-temperature helium. *Entropy* **2021**, *23*, 82. [CrossRef]

68. Albatati, F.; Attar, A. Analytical and experimental study of thermoelectric generator (TEG) system for automotive exhaust waste heat recovery. *Energies* **2021**, *14*, 204. [CrossRef]

69. Feng, H.J.; Wu, Z.X.; Chen, L.G.; Ge, Y.L. Constructal thermodynamic optimization for dual-pressure organic Rankine cycle in waste heat utilization system. *Energy Convers. Manag.* **2021**, *227*, 113585. [CrossRef]

70. Garmejani, H.A.; Hossainpou, S.H. Single and multi-objective optimization of a TEG system for optimum power, cost and second law efficiency using genetic algorithm. *Energy Convers. Manag.* **2021**, *228*, 113658. [CrossRef]

71. Ge, Y.L.; Chen, L.G.; Feng, H.J. Ecological optimization of an irreversible Diesel cycle. *Eur. Phys. J. Plus* **2021**, *136*, 198. [CrossRef]

72. Chen, L.G.; Meng, F.K.; Ge, Y.L.; Feng, H.J.; Xia, S.J. Performance optimization of a class of combined thermoelectric heating devices. *Sci. China Technol. Sci.* **2020**, *63*, 2640–2648. [CrossRef]

73. Sahin, B.; Kodal, A.; Yavuz, H. Efficiency of a Joule-Brayton engine at maximum power density. *J. Phys. D Appl. Phys.* **1995**, *28*, 1309–1313. [CrossRef]

74. Sahin, B.; Kodal, A.; Yavuz, H. Maximum power density analysis of an endoreversible Carnot heat engine. *Energy* **1996**, *21*, 1219–1225. [CrossRef]

75. Chen, L.G.; Zheng, J.L.; Sun, F.R.; Wu, C. Optimum distribution of heat exchanger inventory for power density optimization of an endoreversible closed Brayton cycle. *J. Phys. D Appl. Phys.* **2001**, *34*, 422–427. [CrossRef]

76. Chen, L.G.; Zheng, J.L.; Sun, F.R.; Wu, C. Power density optimization for an irreversible closed Brayton cycle. *Open Syst. Inf. Dyn.* **2001**, *8*, 241–260. [CrossRef]

77. Chen, L.G.; Zheng, J.L.; Sun, F.R.; Wu, C. Performance comparison of an endoreversible closed variable-temperature heat reservoir Brayton cycle under maximum power density and maximum power conditions. *Energy Convers. Manag.* **2002**, *43*, 33–43. [CrossRef]

78. Chen, L.G.; Zheng, J.L.; Sun, F.R.; Wu, C. Performance comparison of an irreversible closed variable-temperature heat reservoir Brayton cycle under maximum power density and maximum power conditions. *Proc. Inst. Mech. Eng. Part. A J. Power Energy* **2005**, *219*, 559–566. [CrossRef]

79. Gonca, G. Thermodynamic analysis and performance maps for the irreversible Dual-Atkinson cycle engine (DACE) with considerations of temperature-dependent specific heats, heat transfer and friction losses. *Energy Convers. Manag.* **2016**, *111*, 205–216. [CrossRef]

80. Gonca, G.; Bahri Sahin, B.; Cakir, M. Performance assessment of a modified power generating cycle based on effective ecological power density and performance coefficient. *Int. J. Exergy* **2020**, *33*, 153–164. [CrossRef]

81. Karakurt, A.S.; Bashan, V.; Ust, Y. Comparative maximum power density analysis of a supercritical CO_2 Brayton power cycle. *J. Therm. Eng.* **2020**, *6*, 50–57. [CrossRef]

82. Angulo-Brown, F. An ecological optimization criterion for finite-time heat engines. *J. Appl. Phys.* **1991**, *69*, 7465–7469. [CrossRef]

83. Yan, Z.J. Comment on "ecological optimization criterion for finite-time heat engines". *Eur. J. Appl. Physiol.* **1993**, *73*, 3583.

84. Cheng, C.Y.; Chen, C.K. Ecological optimization of an endoreversible Brayton cycle. *Energy Convers. Manag.* **1998**, *39*, 33–44. [CrossRef]

85. Ma, Z.S.; Chen, Y.; Wu, J.H. Ecological optimization for a combined diesel-organic Rankine cycle. *AIP Adv.* **2019**, *9*, 015320. [CrossRef]

86. Ahmadi, M.H.; Pourkiaei, S.M.; Ghazvini, M.; Pourfayaz, F. Thermodynamic assessment and optimization of performance of irreversible Atkinson cycle. *Iran. J. Chem. Chem. Eng.* **2020**, *39*, 267–280.

87. Levario-Medina, S.; Valencia-Ortega, G.; Barranco-Jimenez, M.A. Energetic optimization considering a generalization of the ecological criterion in traditional simple-cycle and combined cycle power plants. *J. Non-Equilib. Thermodyn.* **2020**, *45*, 269–290. [CrossRef]

88. Wu, H.; Ge, Y.L.; Chen, L.G.; Feng, H.J. Power, efficiency, ecological function and ecological coefficient of performance optimizations of an irreversible Diesel cycle based on finite piston speed. *Energy* **2021**, *216*, 119235. [CrossRef]

89. Kaushik, S.C.; Tyagi, S.K.; Singhal, M.K. Parametric study of an irreversible regenerative Brayton cycle with isothermal heat addition. *Energy Convers. Manag.* **2003**, *44*, 2013–2025. [CrossRef]

90. Tyagi, S.K.; Kaushik, S.C.; Tiwari, V. Ecological optimization and parametric study of an irreversible regenerative modified Brayton cycle with isothermal heat addition. *Entropy* **2003**, *5*, 377–390. [CrossRef]

91. Tyagi, S.K.; Chen, J. Performance evaluation of an irreversible regenerative modified Brayton heat engine based on the thermoeconomic criterion. *Int. J. Power Energy Syst.* **2006**, *26*, 66–74. [CrossRef]

92. Kumar, R.; Kaushik, S.C.; Kumar, R. Power optimization of an irreversible regenerative Brayton cycle with isothermal heat addition. *J. Therm. Eng.* **2015**, *1*, 279–286. [CrossRef]

93. Tyagi, S.K.; Chen, J.; Kaushik, S.C. Optimum criteria based on the ecological function of an irreversible intercooled regenerative modified Brayton cycle. *Int. J. Exergy* **2005**, *2*, 90–107. [CrossRef]

94. Tyagi, S.K.; Wang, S.; Kaushik, S.C. Irreversible modified complex Brayton cycle under maximum economic condition. *Indian J. Pure Appl. Phys.* **2006**, *44*, 592–601.

95. Tyagi, S.K.; Chen, J.; Kaushik, S.C.; Wu, C. Effects of intercooling on the performance of an irreversible regenerative modified Brayton cycle. *Int. J. Power Energy Syst.* **2007**, *27*, 256–264. [CrossRef]

96. Tyagi, S.K.; Wang, S.; Park, S.R. Performance criteria on different pressure ratios of an irreversible modified complex Brayton cycle. *Indian J. Pure Appl. Phys.* **2008**, *46*, 565–574.

97. Wang, J.H.; Chen, L.G.; Ge, Y.L.; Sun, F.R. Power and power density analyzes of an endoreversible modified variable-temperature reservoir Brayton cycle with isothermal heat addition. *Int. J. Low-Carbon Technol.* **2016**, *11*, 42–53. [CrossRef]

98. Wang, J.H.; Chen, L.G.; Ge, Y.L.; Sun, F.R. Ecological performance analysis of an endoreversible modified Brayton cycle. *Int. J. Sustain. Energy* **2014**, *33*, 619–634. [CrossRef]

99. Tang, C.Q.; Feng, H.J.; Chen, L.G.; Wang, W.H. Power density analysis and multi-objective optimization for a modified endoreversible simple closed Brayton cycle with one isothermal heating process. *Energy Rep.* **2020**, *6*, 1648–1657. [CrossRef]

100. Arora, R.; Kaushik, S.C.; Kumar, R.; Arora, R. Soft computing based multi-objective optimization of Brayton cycle power plant with isothermal heat addition using evolutionary algorithm and decision making. *Appl. Soft Comput.* **2016**, *46*, 267–283. [CrossRef]

101. Arora, R.; Arora, R. Thermodynamic optimization of an irreversible regenerated Brayton heat engine using modified ecological criteria. *J. Therm. Eng.* **2020**, *6*, 28–42. [CrossRef]

102. Chen, L.G.; Tang, C.Q.; Feng, H.J.; Ge, Y.L. Power, efficiency, power density and ecological function optimizations for an irreversible modified closed variable-temperature reservoir regenerative Brayton cycle with one isothermal heating process. *Energies* **2020**, *13*, 5133. [CrossRef]

103. Qi, W.; Wang, W.H.; Chen, L.G. Power and efficiency performance analyses for a closed endoreversible binary Brayton cycle with two isothermal processes. *Therm. Sci. Eng. Prog.* **2018**, *7*, 131–137. [CrossRef]

104. Tang, C.Q.; Chen, L.G.; Feng, H.J.; Wang, W.H.; Ge, Y.L. Power optimization of a closed binary Brayton cycle with isothermal heating processes and coupled to variable-temperature reservoirs. *Energies* **2020**, *13*, 3212. [CrossRef]

105. Ahmadi, M.H.; Dehghani, S.; Mohammadi, A.H.; Feidt, M.; Barranco-Jimenez, M.A. Optimal design of a solar driven heat engine based on thermal and thermo-economic criteria. *Energy Convers. Manag.* **2013**, *75*, 635–642. [CrossRef]

106. Ahmadi, M.H.; Mohammadi, A.H.; Dehghani, S.; Barranco-Jimenez, M.A. Multi-objective thermodynamic-based optimization of output power of Solar Dish-Stirling engine by implementing an evolutionary algorithm. *Energy Convers. Manag.* **2013**, *75*, 438–445. [CrossRef]

107. Ahmadi, M.H.; Ahmadi, M.A.; Mohammadi, A.H.; Feidt, M.; Pourkiaei, S.M. Multi-objective optimization of an irreversible Stirling cryogenic refrigerator cycle. *Energy Convers. Manag.* **2014**, *82*, 351–360. [CrossRef]

108. Ahmadi, M.H.; Ahmadi, M.A.; Mehrpooya, M.; Hosseinzade, H.; Feidt, M. Thermodynamic and thermo-economic analysis and optimization of performance of irreversible four- temperature-level absorption refrigeration. *Energy Convers. Manag.* **2014**, *88*, 1051–1059. [CrossRef]

109. Ahmadi, M.H.; Ahmadi, M.A. Thermodynamic analysis and optimization of an irreversible Ericsson cryogenic refrigerator cycle. *Energy Convers. Manag.* **2015**, *89*, 147–155. [CrossRef]

110. Jokar, M.A.; Ahmadi, M.H.; Sharifpur, M.; Meyer, J.P.; Pourfayaz, F.; Ming, T.Z. Thermodynamic evaluation and multi-objective optimization of molten carbonate fuel cell-supercritical CO2 Brayton cycle hybrid system. *Energy Convers. Manag.* **2017**, *153*, 538–556. [CrossRef]
111. Han, Z.H.; Mei, Z.K.; Li, P. Multi-objective optimization and sensitivity analysis of an organic Rankine cycle coupled with a one-dimensional radial-inflow turbine efficiency prediction model. *Energy Convers. Manag.* **2018**, *166*, 37–47. [CrossRef]
112. Ghasemkhani, A.; Farahat, S.; Naserian, M.M. Multi-objective optimization and decision making of endoreversible combined cycles with consideration of different heat exchangers by finite time thermodynamics. *Energy Convers. Manag.* **2018**, *171*, 1052–1062. [CrossRef]
113. Ahmadi, M.H.; Jokar, M.A.; Ming, T.Z.; Feidt, M.; Pourfayaz, F.; Astaraei, F.R. Multi-objective performance optimization of irreversible molten carbonate fuel cell–Braysson heat engine and thermodynamic analysis with ecological objective approach. *Energy* **2018**, *144*, 707–722. [CrossRef]
114. Wang, M.; Jing, R.; Zhang, H.R.; Meng, C.; Li, N.; Zhao, Y.R. An innovative Organic Rankine Cycle (ORC) based Ocean Thermal Energy Conversion (OTEC) system with performance simulation and multi-objective optimization. *Appl. Therm. Eng.* **2018**, *145*, 743–754. [CrossRef]
115. Patela, V.K.; Raja, B.D. A comparative performance evaluation of the reversed Brayton cycle operated heat pump based on thermo-ecological criteria through many and multi-objective approaches. *Energy Convers. Manag.* **2019**, *183*, 252–265. [CrossRef]
116. Hu, S.Z.; Li, J.; Yang, F.B.; Yang, Z.; Duan, Y.Y. Multi-objective optimization of organic Rankine cycle using hydrofluorolefins (HFOs) based on different target preferences. *Energy* **2020**, *203*, 117848. [CrossRef]
117. Hu, S.Z.; Li, J.; Yang, F.B.; Yang, Z.; Duan, Y.Y. How to design organic Rankine cycle system under fluctuating ambient temperature: A multi-objective approach. *Energy Convers. Manag.* **2020**, *224*, 113331. [CrossRef]
118. Sun, M.; Xia, S.J.; Chen, L.G.; Wang, C.; Tang, C.Q. Minimum entropy generation rate and maximum yield optimization of sulfuric acid decomposition process using NSGA-II. *Entropy* **2020**, *22*, 1065. [CrossRef] [PubMed]
119. Sadeghi, S.; Ghandehariun, S.; Naterer, G.F. Exergoeconomic and multi-objective optimization of a solar thermochemical hydrogen production plant with heat recovery. *Energy Convers. Manag.* **2020**, *225*, 113441. [CrossRef]
120. Wu, Z.X.; Feng, H.J.; Chen, L.G.; Ge, Y.L. Performance optimization of a condenser in ocean thermal energy conversion (OTEC) system based on constructal theory and multi-objective genetic algorithm. *Entropy* **2020**, *22*, 641. [CrossRef]
121. Ghorani, M.M.; Haghighi, M.H.S.; Riasi, A. Entropy generation minimization of a pump running in reverse mode based on surrogate models and NSGA-II. *Int. Commun. Heat Mass Transfer* **2020**, *118*, 104898. [CrossRef]
122. Wang, L.B.; Bu, X.B.; Li, H.S. Multi-objective optimization and off-design evaluation of organic Rankine cycle (ORC) for low-grade waste heat recovery. *Energy* **2020**, *203*, 117809. [CrossRef]
123. Herrera-Orozco, I.; Valencia-Ochoa, G.; Jorge Duarte-Forero, J. Exergo-environmental assessment and multi-objective optimization of waste heat recovery systems based on Organic Rankine cycle configurations. *J. Clean. Prod.* **2021**, *288*, 125679. [CrossRef]
124. Shi, S.S.; Ge, Y.L.; Chen, L.G.; Feng, F.J. Four objective optimization of irreversible Atkinson cycle based on NSGA-II. *Entropy* **2020**, *22*, 1150. [CrossRef]
125. Tang, W.; Feng, H.J.; Chen, L.G.; Xie, Z.J.; Shi, J.C. Constructal design for a boiler economizer. *Energy* **2021**, *223*, 120013. [CrossRef]
126. Chen, L.G.; Zheng, J.L.; Sun, F.R.; Wu, C. Power density optimization for an irreversible regenerated closed Brayton cycle. *Phys. Scripta* **2001**, *64*, 184–191. [CrossRef]
127. Bejan, A. *Entropy Generation through Heat and Fluid Flow*; Wiley: New York, NY, USA, 1982.
128. Bejan, A. Theory of heat transfer-irreversible power plant. *Int. J. Heat Mass Transfer* **1988**, *31*, 1211–1219. [CrossRef]
129. Bejan, A. The equivalence of maximum power and minimum entropy generation rate in the optimization of power plants. *J. Energy Res. Tech.* **1996**, *118*, 98–101. [CrossRef]
130. Bejan, A. Models of power plants that generate minimum entropy while operating at maximum power. *Am. J. Phys.* **1996**, *64*, 1054–1059. [CrossRef]
131. Salamon, P.; Hoffmann, K.H.; Schubert, S.; Berry, R.S.; Andresen, B. What conditions make minimum entropy production equivalent to maximum power production? *J. Non-Equilib. Thermodyn.* **2001**, *26*, 73–83. [CrossRef]
132. Andresen, B.; Berry, R.S.; Nitzan, A.; Salamon, P. Thermodynamics in finite time: The step-Carnot cycle. *Phys. Rev. A* **1977**, *15*, 2086–2093. [CrossRef]
133. Orlov, V.N.; Rudenko, A.V. Optimal control in problems of extremal of irreversible thermodynamic processes. *Avtomatika Telemekhanika* **1985**, *46*, 549–577.
134. Lu, P.C. Thermodynamics with finite heat-transfer area or finite surface thermodynamics. Thermodynamics and the Design, Analysis, and Improvement of Energy Systems, ASME Adv. *Energy Sys. Div. Pub. AES* **1995**, *35*, 51–60.
135. Bejan, A. Entropy generation minimization: The new thermodynamics of finite size devices and finite time processes. *J. Appl. Phys.* **1996**, *79*, 1191–1218. [CrossRef]
136. Feidt, M. *Thermodynamique et Optimisation Energetique des Systems et Procedes*, 2nd ed.; Technique et Documentation, Lavoisier: Paris, France, 1996. (In French)
137. Dong, Y.; El-Bakkali, A.; Feidt, M.; Descombes, G.; Perilhon, C. Association of finite-dimension thermodynamics and a bond-graph approach for modeling an irreversible heat engine. *Entropy* **2012**, *14*, 1234–1258. [CrossRef]
138. Feidt, M. *Thermodynamique Optimale en Dimensions Physiques Finies*; Hermès: Paris, France, 2013.

139. Perescu, S.; Costea, M.; Feidt, M.; Ganea, I.; Boriaru, N. *Advanced Thermodynamics of Irreversible Processes with Finite Speed and Finite Dimensions*; Editura AGIR: Bucharest, Romania, 2015.
140. Feidt, M. *Finite Physical Dimensions Optimal Thermodynamics 1. Fundamental*; ISTE Press and Elsevier: London, UK, 2017.
141. Feidt, M. *Finite Physical Dimensions Optimal Thermodynamics 2. Complex. Systems*; ISTE Press and Elsevier: London, UK, 2018.
142. Blaise, M.; Feidt, M.; Maillet, D. Influence of the working fluid properties on optimized power of an irreversible finite dimensions Carnot engine. *Energy Convers. Manag.* **2018**, *163*, 444–456. [CrossRef]
143. Feidt, M.; Costea, M. From finite time to finite physical dimensions thermodynamics: The Carnot engine and Onsager's relations revisited. *J. Non-Equilib. Thermodyn.* **2018**, *43*, 151–162. [CrossRef]
144. Dumitrascu, G.; Feidt, M.; Popescu, A.; Grigorean, S. Endoreversible trigeneration cycle design based on finite physical dimensions thermodynamics. *Energies* **2019**, *12*, 3165. [CrossRef]
145. Feidt, M.; Costea, M. Progress in Carnot and Chambadal modeling of thermomechnical engine by considering entropt and heat transfer entropy. *Entropy* **2019**, *21*, 1232. [CrossRef]
146. Feidt, M.; Costea, M.; Feidt, R.; Dancl, Q.; Périlhon, C. New criteria to characterize the waste heat recovery. *Energies* **2020**, *13*, 789. [CrossRef]

Article

Stirling Refrigerating Machine Modeling Using Schmidt and Finite Physical Dimensions Thermodynamic Models: A Comparison with Experiments

Cătălina Dobre [1], Lavinia Grosu [2], Alexandru Dobrovicescu [1], Georgiana Chişiu [3] and Mihaela Constantin [1,*]

[1] Department of Engineering Thermodynamics, Engines, Thermal and Refrigeration Equipment, University Politehnica of Bucharest, Splaiul Independenței 313, 060042 Bucharest, Romania; catalina.dobre@upb.ro (C.D.); adobrovicescu@yahoo.com (A.D.)

[2] Laboratory of Energy, Mechanics and Electromagnetic, Paris West Nanterre La Défense University, 50, Rue de Sèvres, 92410 Ville d'Avray, France; mgrosu@parisnanterre.fr

[3] Department of Machine Elements and Tribology, University Politehnica of Bucharest, Splaiul Independenței 313, 060042 Bucharest, Romania; georgiana.chisiu@upb.ro

* Correspondence: i.mihaelaconstantin@gmail.com

Abstract: The purpose of the study is to show that two simple models that take into account only the irreversibility due to temperature difference in the heat exchangers and imperfect regeneration are able to indicate refrigerating machine behavior. In the present paper, the finite physical dimensions thermodynamics (FPDT) method and 0-D modeling using the Schmidt model with imperfect regeneration were applied in the study of a β type Stirling refrigeration machine. The 0-D modeling is improved by including the irreversibility caused by imperfect regeneration and the finite temperature difference between the gas and the heat exchangers wall. A flowchart of the Stirling refrigerator exergy balance is presented to show the internal and external irreversibilities. It is found that the irreversibility at the regenerator level is more important than that at the heat exchangers level. The energies exchanged by the working gas are expressed according to the practical parameters, necessary for the engineer during the entire project. The results of the two thermodynamic models are presented in comparison with the experimental results, which leads to validation of the proposed FPDT model for the functional and constructive parameters of the studied refrigerating machine.

Keywords: Stirling refrigerator; thermodynamic analysis; numerical model; imperfect regeneration

Citation: Dobre, C.; Grosu, L.; Dobrovicescu, A.; Chişiu, G.; Constantin, M. Stirling Refrigerating Machine Modeling Using Schmidt and Finite Physical Dimensions Thermodynamic Models: A Comparison with Experiments. *Entropy* **2021**, *23*, 368. https://doi.org/10.3390/e23030368

Academic Editor: T M Indra Mahlia

Received: 27 January 2021
Accepted: 17 March 2021
Published: 19 March 2021

Publisher's Note: MDPI stays neutral with regard to jurisdictional claims in published maps and institutional affiliations.

1. Introduction

The continued growth in the demand for refrigeration in almost all parts of the world and global warming due to the consumption of chlorofluorocarbon (HCFC) refrigerant has led the engineering community to seek applications for vapor-compression refrigeration. The Stirling refrigeration cycle is an important cycle model in the research and manufacture of refrigerators. The Stirling cycle machine is an alternative that could work with an environmentally friendly cooling fluid [1].

The Stirling cycle refrigerating machine was first developed in 1832 [2] but the system was first practically made in 1862, when Alexander Kirk built and patented a closed-cycle refrigerator. In 1971, Beale stated that by reversing the cycle, the Stirling cycle could be used for both work production and refrigeration purposes [3].

The Stirling reversible refrigeration cycle, for the same temperature range under perfect regenerative conditions [4], has the same coefficient of performance as the Carnot reversible refrigeration cycle according to classical thermodynamics.

The Stirling refrigerator is composed of two chambers with variable volume (expansion and compression) physically separated from the regenerator and with different temperatures. The presence of the regenerator (an economizer) qualifies the Stirling cycle machine as a regenerative machine.

105

According to the classical theory of thermodynamics, the performance of a Stirling cycle machine is a function of pressure, the ratio between temperature, speed and phase angle, fluid type, the efficiency of heat exchangers and volume [1].

Various thermodynamic models of Stirling machine operations have been proposed in the literature, with various assumptions. Schmidt developed the first performance analysis of the Stirling machine in 1871 and this proved to be an effective aspect of its design [5]. After Curzon and Ahlborn [6] studied the Carnot direct cycle using finite heat transfer, models were developed using finite time thermodynamics (FTT) [7], finite physical dimensions thermodynamics (FPDT) [8,9], and finite speed thermodynamics (FST) [10], which have been applied to several types of machines, including Stirling machines.

A finite time heat transfer analysis [7] was performed in 1998 for an air refrigeration cycle with non-isentropic compression and expansion. The relation between the coefficient of performance (COP) and the cooling load with the pressure ratio was obtained.

Petrescu et al. [10] developed an analytical model for estimating the performance of a Stirling engine based on the first and second laws of thermodynamics, called finite speed thermodynamics (FST). The model [11] directly connects the irreversibilities, and the flow and mechanical friction are taken into account.

Chen [12] developed an irreversible cycle model in order to predict the performance and input power required for a Stirling refrigerator optimized to a specified cooling capacity.

A β-type Stirling cycle refrigeration machine was mathematically designed and experimentally tested in [13]. Those authors studied the types of working fluids, the effect of the phase difference of the piston and the displacer on the refrigeration performance, the effect of parameters such as the ratio between the expansion volume and the compression volume and the dead volume ratio.

For the Stirling cycle refrigerator, Ataer and Karabulut [14] performed an analysis on the thermodynamic control volume subjected to periodic mass flow and evaluated the performed activity, instantaneous pressure and coefficient of performance.

A nonlinear mathematical model was developed for an air-filled Stirling alpha refrigerator by incorporating thermodynamics, wall heat transfer and fluid resistance in the regenerator. Different variables were also determined for both workspaces [15].

The effects of different parameters on the cooling performance of a Stirling cryocooler were also investigated [16]. It was found that the highest work loss was due to mechanical friction loss and the highest heat loss was due to conduction loss.

The performance of a β-type Stirling refrigeration machine with a regenerative displacer was studied by Hachem et al. [17,18], considering the complex phenomena related to the mechanics of compressible fluids, heat transfer and thermodynamics for energy analysis. An experimental validation with a focus on evaluating the effect of geometric parameters, such as the expansion space, the volume of the dead space and the compression of the swept volume was performed in [18]. The authors analyzed and optimized the parameters of the regenerator regarding the performance of the refrigerator. The various losses associated with the Stirling refrigerator that directly affect its cooling performance were evaluated. They described these losses as a function of the length and diameter of the regenerator.

Given the imperfection of the practical regenerator, researchers [19,20] have developed many thermodynamic models of Stirling engines using finite time thermodynamics (FTT). Based on finite speed thermodynamics (FST), Petrescu et al. [21] developed a method for calculating the coefficient that characterizes regenerative loss in a Stirling machine, based on the first law for processes with finite speed. Based on isothermal theory, Kongtragool and Formosa [22] studied the effect of regenerative efficiency and dead volume on a Stirling engine with an imperfect regenerator.

In the present paper, a finite physical dimensions thermodynamic (FPDT) method and 0-D modeling (isothermal analysis) using the Schmidt model (second order) with imperfect regeneration were applied in the study of a β-type Stirling refrigeration machine, with academic use and benefit.

The findings of Feidt et al. [23] show that the most significant reduction in performance is due to the non-adiabatic regenerator. The isotherm model in this paper is improved by including the irreversibility caused by imperfect regeneration and the finite temperature difference between the gas and the wall of the heat exchangers (cold and hot). The numerical model describes the evolution of instantaneous variables (pressure, volume, mass, changed energy, irreversibility) depending on the rotation angle of the shaft.

The FPDT model [16,24] is based on the irreversible thermodynamics approach, which is an old approach, but has had some improvements and engineering adjustments, which were the aims of recent papers by Grosu et al. [25–27].

The results obtained after applying the two models of thermodynamic analysis justify a more realistic evaluation of the FPDT model by reporting the experimental results. In this context, in order to identify the limitations of the isothermal model, this research was completed with an exergetic analysis of a β-type Stirling refrigerator that allows the development of a system of equations that describes the processes that take place at each element of the machine. The purpose of developing this method of thermodynamic analysis was to establish the value of irreversible losses in the actual cycle of the refrigeration machine and determine the cycle component to be improved in order to reduce the degree of irreversibility of the cycle.

2. Materials and Method

2.1. Description of Experimental Installation

A β-type Stirling refrigerator consisting of an arrangement with a displacer, power piston and regenerator in line was analyzed. A cylinder of highly resilient glass is surrounded by a water jacket in which a stream of water is the hot tank of the system operating as a refrigeration machine. The displacer forces the gas (air) to pass from the bottom space to the top space of the cylinder and vice versa. It also has an extremely conductive material, which is used for heat storage/release, thus acting as a regenerator, in order to improve the efficiency. The two pistons perform an alternating reciprocating motion with an angle of 110°.

The experimental device can function as an engine by providing mechanical work, or as a refrigerating machine (reverse cycle) by using an electric motor that drives the machine shaft [28]. The configuration of the Stirling refrigerator proposed for this study is shown in Figure 1. At the top of the cylinder is a thermocouple that allows temperature measurement and an electrical resistance that helps to determine the refrigerating power through a compensation method.

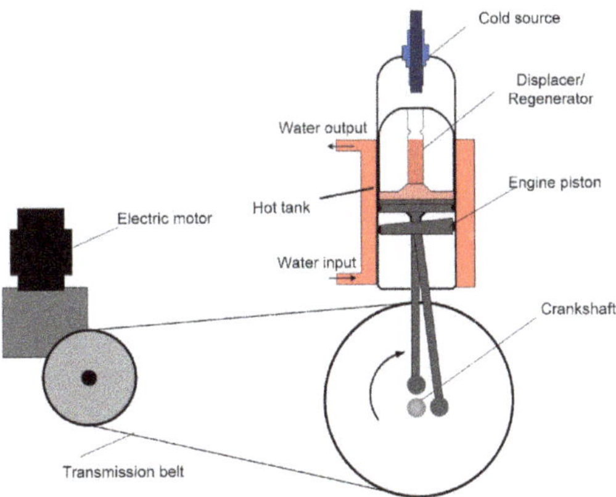

Figure 1. Experimental device using a β-type Stirling refrigerating machine.

2.2. Application of Schmidt Method with Imperfect Regeneration in the Study of a β-Type Stirling Refrigerating Machine

The isothermal analysis (Schmidt method) takes into account the external and internal irreversibility of the machine and the kinematics of the pistons. In addition, the uneven distribution of time and space of the working fluid in the machine is taken into consideration by dividing the refrigerator into three volumes associated with a characteristic temperature. The assumptions that Schmidt considered in his analysis included the following: (a) the fluid in the compression volume of the refrigerating machine and the cold exchanger is always kept at a constant temperature, and the fluid temperature in the expansion volume and the hot-end heat exchanger is constant; (b) the surface temperature of the cylinder and the piston is constant; (c) the mass of the fluid is constant, which implies that there is no leakage and the same instantaneous pressure on the whole machine; (d) an ideal gas is used as a working fluid (perfect gas equation of state is applied); (e) there is harmonic/sinusoidal movement of the pistons (idealized crankshaft); and (f) the speed of working fluid within the machine is constant. The hypothesis of energy loss independence is used in this method [28].

In practice, this hypothesis, according to which the gas behaves isothermally in the expansion and compression spaces, is not true at high speeds. At high speeds, compression and expansion processes are closer to adiabatic processes [5].

Given the constructive peculiarities of the machine studied in this paper, it was operated at very low speeds and in order to model the Stirling refrigeration machine with some realism, the isothermal model was adapted.

The Schmidt method is based on dividing the refrigeration machine into three spaces: the expansion volume, the regenerator volume, and the compression volume. (Figure 2). Each part is considered a control volume, to which the laws of energy and mass conservation are applied.

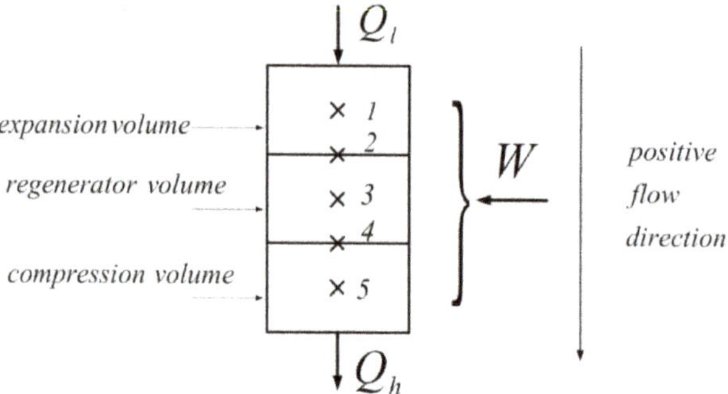

Figure 2. Representation of three volumes of machine and their boundaries.

According to the assumption, the gas temperature history will remain the same and part of the regenerated heat loss will be continuously compensated by a heat supplement $Q_{p,reg}$ provided by the source, as each cycle is driven by imperfect regeneration (Figure 3). Using refrigerator geometry, the volumes of compression and expansion spaces can be expressed according to the instantaneous positions of the pistons [29].

The following equation is used to determine the instantaneous volume of the compression space (hot):

$$V_C = \frac{V_{C0}}{2}[1 - \cos \varphi] + V_{mC},\qquad(1)$$

where φ is the idealized crankshaft rotation angle and V_{C0} is the swept compression volume; this is the displacer swept volume in the case of β-type Stirling machines.

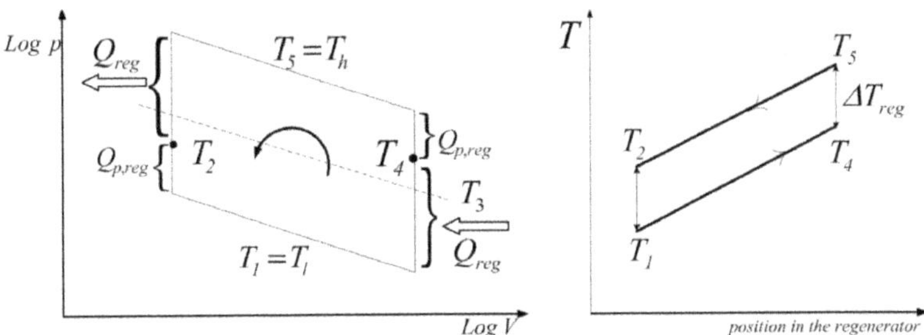

Figure 3. Temperature gradient in refrigerator regenerator.

The instantaneous volume of the expansion space (cold) is a combination of several volumes and can be determined as:

$$V_E = \left\{ \frac{V_{C0}}{2} \cdot [1 + \cos(\varphi)] + \frac{V_{E0}}{2}[1 - \cos(\varphi - \varphi_0)] - V_{0l} \right\} + V_{mE}, \qquad (2)$$

where φ_0 is the phase lag angle of the piston movements and V_{E0} is the swept expansion volume. V_{0l} is the overlapping volume in the case of a β-type Stirling machine and is due to the intrusion of the displacer piston into the working piston swept volume.

The dead volumes V_{mE} and V_{mC} on the heat exchangers are also taken into account.

To evaluate the mass of fluid in each volume, the state equation of the perfect gases is used. The instantaneous pressure is assumed to be uniform in the machine and its variation can be determined by using the mass balance:

$$p = \frac{mR}{\frac{V_h}{T_h} + \frac{V_{reg}}{T_r} + \frac{V_l}{T_l}} \qquad (3)$$

The elementary masses of each volume are calculated with:

$$\begin{cases} dm_l = \frac{pdV_l + V_l dp}{RT_1} = dm_1 \\ dm_h = \frac{pdV_h + V_h dp}{RT_5} = dm_5 \\ dm_{reg} = m_{reg}\frac{dp}{p} = dm_5 \end{cases} \qquad (4)$$

Considering the mass flow direction on the interface, the interface temperatures can be expressed as follows:

$dm_2 = -dm_l$ if $dm_2 < 0$, then $T_2 = T_1 + \Delta T_{reg}$, otherwise, $T_2 = T_1$;

$dm_4 = -dm_h$ if $dm_4 < 0$, then $T_4 = T_5$, otherwise, $T_4 = T_5 - \Delta T_{reg}$.

While differentiating Equation (3) and considering that the temperatures are constant, dp is obtained in the following form:

$$dp = \frac{-p\left(\frac{dV_l}{T_1} + \frac{dV_h}{T_5} \right)}{\frac{V_l}{T_1} + \frac{V_{reg}}{T_3} + \frac{V_h}{T_5}} \qquad (5)$$

The internal irreversibility of the studied Stirling cycle is assumed to be due to the imperfect regeneration. The regenerator/displacer reciprocating movement forces the air of the cooling space toward the heating space and conversely: it is also useful to store and release the heat exchanged with the regenerator material during this transfer (Figure 3). The difference is the temperature gap on the regenerator ΔT_{reg}, assumed to be constant on the whole length of the regenerator [28]. Therefore, in the case of the Stirling refrigerator, the regenerator efficiency is defined by:

$$\eta_{reg} = \frac{T_5 - T_4}{T_5 - T_1} = \frac{T_2 - T_1}{T_5 - T_1} = \frac{\Delta T_{reg}}{T_5 - T_1} \tag{6}$$

Thus:

$$\begin{aligned} T_5 &= T_4 + \eta_{reg}(T_5 - T_1) \\ T_2 &= T_1 + \eta_{reg}(T_5 - T_1) \end{aligned} \tag{7}$$

In the regenerator, the changed work is zero and the average temperature is supposed to be constant T_{reg}. The regenerator temperature, $T_{reg} = T_3$, is a logarithmic average of cold and hot space (V_C and V_E) temperatures:

$$T_3 = T_{reg} = \frac{T_h - T_l}{\ln \frac{T_h}{T_l}} = \frac{T_5 - T_1}{\ln \frac{T_5}{T_1}} \tag{8}$$

The quantity of heat changed at the level of the three volumes is obtained starting from the energy conservation equation applied to each volume:

$$\begin{cases} \delta Q_l = \left(\frac{c_p}{R} + 1\right) p dV_l + \frac{c_p}{R} V_l dp + c_p T_2 dm_2 \\ \delta Q_{reg} = V_{reg} \frac{c_p}{R} dp + c_p(T_4 dm_4 - T_2 dm_2) \\ \delta Q_h = \left(\frac{c_p}{R} + 1\right) p dV_h + \frac{c_p}{R} V_h dp - c_p T_4 dm_4 \end{cases} \tag{9}$$

The elementary mechanical work in the compression $\delta W_h = -pdV_h$ and expansion $\delta W_l = -pdV_l$ spaces allow, after integration, calculation of the mechanical work consumed in a cycle:

$$W = W_l + W_h \tag{10}$$

The temperatures of the expansion and compression spaces are determined starting with heat flow rates and from the global heat transfer coefficients, experimentally obtained:

$$\left|\dot{Q}_h\right| = hA_h(T_h - T_{wh}) \rightarrow T_h = T_{wh} + \frac{\left|\dot{Q}_h\right|}{hA_h} \tag{11}$$

$$\dot{Q}_l = hA_l(T_{wl} - T_l) \rightarrow T_l = T_{wl} - \frac{\dot{Q}_l}{hA_l} \tag{12}$$

The equations presented above were solved using the Simulink simulation tool. In order to improve the obtained results based on the isothermal method (Schmidt) by taking temperature levels into account, an exergetic analysis is required for the β-type Stirling refrigerating machine.

2.3. Application of Exergetic Method in the Study of β-Type Stirling Refrigerating Machine
2.3.1. Exergetic Analysis Applied in the Study of the β-Type Stirling Refrigeration Machine

The simple and fast processing of the energy balance and exergetic balance equations leads to obtaining the classic exergetic balance equations customized on the reversed cycle (refrigeration installation), written at the consumer level:

$$\left|\dot{W}\right| = \left|\dot{Ex}_{Q_{wl}}^{T_{wl}}\right| + \left|\dot{Ex}_{Q_h}^{T_h}\right| + \dot{Ex}_l^D + \dot{Ex}_{reg}^D, \tag{13}$$

where:
$\dot{Ex}_l^D$ is exergy destruction due to heat transfer at the finite difference in the cooler; $\dot{Ex}_{reg}^D$ is the exergy destruction in the regenerator and $\left|\dot{Ex}_{Q_h}^{T_h}\right|$ is loss of exergy with heat discharged into the environment.

The schematic of the exergy balance for the Stirling refrigerating machine is presented in Figure 4.

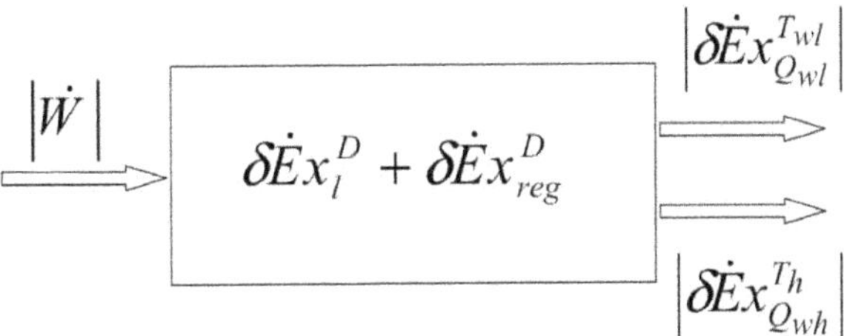

Figure 4. Exergy balance for β-type Stirling refrigerating machine.

For the calculation of terms in Equation (13), in the following, the exergetic balances are established at the level of each element of the Stirling machine, depending on the kinematics of the pistons.

The exergetic balance in differential form is applied for each heat exchanger:

$$dEx = \delta Ex_Q^T + \delta W + p_0 dV + ex_i^f dm_i - ex_e^f dm_e - T_0 \delta \Pi, \tag{14}$$

where δEx_Q^T is the exergy of heat at the temperature T of the system.

Then we obtain:

$$dEx_{reg} = \left(1 - \frac{T_0}{T_{reg}}\right)\delta Q_{reg} + \delta W_{reg} + p_0 dV_{reg} + ex_2^f dm_2 - ex_4^f dm_4 - T_0 \delta \Pi_{reg} \tag{15}$$

$$dEx_l = \left(1 - \frac{T_0}{T_l}\right)\delta Q_l + \delta W_l + p_0 dV_l - ex_2^f dm_2 \tag{16}$$

$$dEx_h = \left(1 - \frac{T_0}{T_h}\right)\delta Q_h + \delta W_h + p_0 dV_h + ex_4^f dm_4 \tag{17}$$

For a cycle, the balance can be written as follows:

$$dEx_l + dEx_{reg} + dEx_h = 0 \tag{18}$$

Using the equations presented above, $\delta \Pi_{reg}$ and Π_{reg} can also be calculated.

2.3.2. Study of Heat Exchangers (Compression and Expansion Volume) and Calculation of Exergy Destroyed due to Temperature Differences
Cold-End Heat Exchanger Study

A functional diagram of the expansion volume (Figure 5) shows the entropies and exergies exchanged by the air in the expansion volume of the refrigeration machine with a cold source (cylinder head).

The exergetic balance allows the determination of exergy lost due to temperature differences between the expansion volume and the cylinder head:

$$\left|\delta Ex_{Q_l}^{T_l}\right| = \delta Ex_l^D + \delta Ex_{Q_{wl}}^{T_{wl}}, \tag{19}$$

where:

$$\delta Ex_{Q_l}^{T_l} = \left(1 - \frac{T_0}{T_l}\right)\delta Q_l < 0 \tag{20}$$

represents the exergy of heat δQ_l at temperature T_l.

The exergy of heat δQ_{wl} at temperature T_{wl} is given by the equation:

$$\delta Ex_{Q_{wl}}^{T_{wl}} = \left(1 - \frac{T_0}{T_{wl}}\right)\delta Q_{wl} > 0,$$

(21)

where $\delta Ex_{Q_{wl}}^{T_{wl}}$ is the useful effect of the refrigeration machine in exergetic terms.

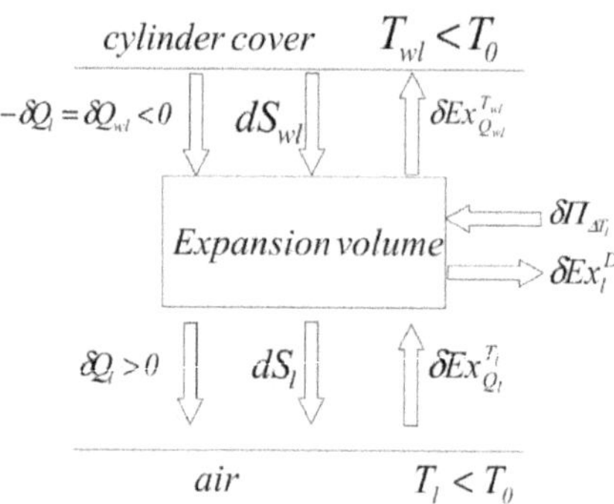

Figure 5. Exergetic and entropic functional diagram of expansion volume.

Replacing relations (20) and (21) in Equation (19) of the exegetical balance, the destroyed exergy at the level of the cold exchanger results in:

$$\delta Ex_l^D = \left|\delta Ex_{Q_l}^{T_l}\right| - \delta Ex_{Q_{wl}}^{T_{wl}} = \left(\frac{T_0}{T_l} - 1\right)\delta Q_l - \left(1 - \frac{T_0}{T_{wl}}\right)(-\delta Q_l)$$

(22)

By grouping the terms, we can obtain:

$$\delta Ex_l^D = T_0 \delta Q_l \left(\frac{1}{T_l} - \frac{1}{T_{wl}}\right)$$

(23)

The destroyed exergy flow rate is calculated by integrating relation (23) over the entire cycle of the refrigeration machine during a complete rotation of the shaft:

$$\dot{Ex}_l^D = n \oint \delta Ex_l^D$$

(24)

The exergetic efficiency of the cold exchanger is:

$$\eta_{ex_l} = \frac{\dot{Ex}_{Q_{wl}}^{T_{wl}}}{\left|\dot{Ex}_{Q_l}^{T_l}\right|},$$

(25)

and the dissipation coefficient is:

$$\zeta_l = \frac{\dot{Ex}_l^D}{\left|\dot{Ex}_{Q_l}^{T_l}\right|}$$

(26)

Hot-End Heat Exchanger Study

The gas temperature of the compression volume is higher than the ambient temperature, and as the gas is cooled, its exergy will decrease, so the exergy flow rate will have the same direction as the heat transfer.

A functional diagram of the compression volume shows the exchanged air exergy with a cold source (Figure 6).

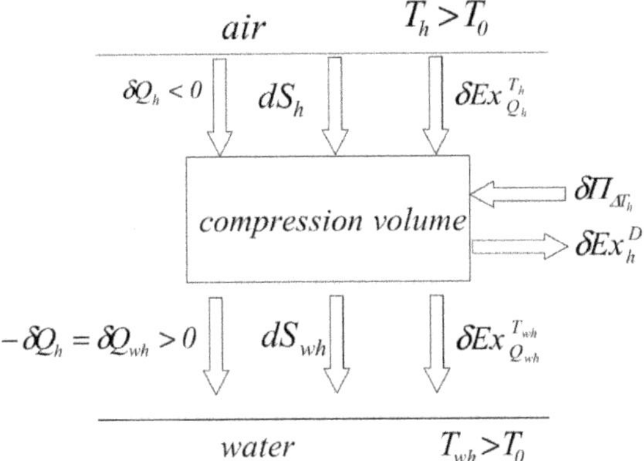

Figure 6. Exergetic and entropic functional diagram of compression volume.

The exergetic balance of the compression chamber can be written as follows:

$$\left| \delta Ex_{Q_h}^{T_h} \right| = \delta Ex_h^D + \delta Ex_{Q_{wh}}^{T_{wh}}, \tag{27}$$

where the exergy of heat δQ_h at temperature T_h is:

$$\delta Ex_{Q_h}^{T_h} = \left(1 - \frac{T_0}{T_h} \right) \delta Q_h < 0 \tag{28}$$

The exergy of heat δQ_{wh} at temperature T_{wh} is:

$$\delta Ex_{Q_{wh}}^{T_{wh}} = \left(1 - \frac{T_0}{T_{wh}} \right) \delta Q_{wh} > 0 \tag{29}$$

The exergy lost at the hot-end heat exchanger due to the temperature difference between the compression room and the hot source can be calculated as follows:

$$\delta Ex_h^D = \left| \delta Ex_{Q_h}^{T_h} \right| - \delta Ex_{Q_{wh}}^{T_{wh}} = -\left(1 - \frac{T_0}{T_h} \right) \delta Q_h + \left(1 - \frac{T_0}{T_{wh}} \right) \delta Q_h \tag{30}$$

$$\delta Ex_h^D = T_0 \delta Q_h \left(\frac{1}{T_h} - \frac{1}{T_{wh}} \right), \tag{31}$$

and the exergy destroyed at the hot-end heat exchanger level is:

$$\dot{Ex}_h^D = n \oint \delta Ex_h^D \tag{32}$$

The exergetic efficiency of the hot-end heat exchanger can be calculated as:

$$\eta_{ex_h} = \frac{\dot{Ex}_{Q_{wh}}^{T_{wh}}}{\left| \dot{Ex}_{Q_h}^{T_h} \right|}, \tag{33}$$

and its dissipation coefficient as:

$$\varsigma_h = \frac{\dot{Ex}_h^D}{\left| \dot{Ex}_{Q_h}^{T_h} \right|} \tag{34}$$

2.4. Application of TDFF in the Study of the β-Type Stirling Refrigerating Machine

Finite physical dimensions thermodynamics (FPDT) [13–17] is a method that regroups the techniques of thermodynamics in finite time, speed and geometric dimensions. This method introduces the exo-irreversibilities due to the finite heat transfer between sources (hot source, cold source, regenerator) and the working fluid. In addition, it considers the constraints faced by engineers. Using classical thermodynamics, it has been shown that machines with or without heat generation operating after cycles similar to the Carnot cycle can be described by using physical parameters such as p_{max}, V_{max}, T_h, and T_l as reference parameters. It is essential to consider the rotation speed as the main variable, because heat and mass transfer are dependent in a straightforward manner on speed and naturally must be expressed accordingly.

In the following, the FPDT method is applied in the study of the exo-irreversible reversed Stirling cycle with imperfect regeneration, represented in Figure 7.

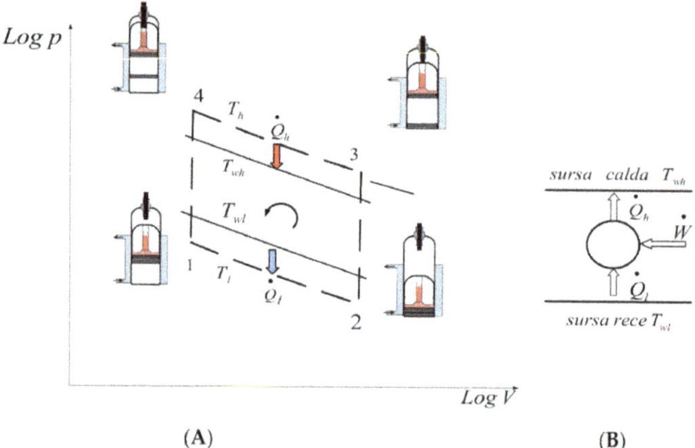

Figure 7. Exo-irreversible reversed Stirling cycle. (**A**) Logp-LogV diagram in the range limit of p_{max}, V_{max}, T_l and T_l; (**B**) energy balance scheme.

The main hypothesis of this method of thermodynamic analysis is that the reheater and the compression space are at the same temperature, as are the cooler and the expansion space.

It is also considered that the gas that is used is a perfect gas and its total mass is supposed to be transferred entirely from the hot volume to the cold volume and vice versa (neglecting the dead volume), remaining constant throughout the experiment (it is considered a closed thermodynamic system).

The energies transferred in the cycle are given by the following relations.

The heat given to the hot tank (water) by the working gas at temperature T_h, in the case of perfect regeneration, in absolute value, is:

$$|Q_{h.rev}| = |Q_{34}| = p_{max} V_{max} \frac{\ln \varepsilon}{\varepsilon} = E_\varepsilon, \tag{35}$$

where E_ε is the reference energy of the FPDT model.

The heat taken from the cold tank by the working gas at temperature T_l, in the case of perfect regeneration, is:

$$Q_{l.rev} = Q_{12} = p_{max} V_{max} \frac{\ln \varepsilon}{\varepsilon} \frac{T_l}{T_h} = E_\varepsilon \frac{T_l}{T_h} \tag{36}$$

The heat exchanged with the regenerator (stored and detached) during an isochoric transformation is:

$$Q_{reg} = mc_v(T_h - T_l) = m\frac{R}{\gamma - 1}T_h\left(1 - \frac{T_l}{T_h}\right) \tag{37}$$

The regeneration efficiency is described by the relation:

$$\eta_{reg} = \frac{Q_{reg} - Q_{p,reg}}{Q_{reg}}, \tag{38}$$

where $Q_{p,reg}$ is the amount of heat to be added to that received by the hot source and given by the cold source ($Q_{p,reg} > 0$).

It follows that:

$$Q_{p,reg} = (1 - \eta_{reg})Q_{reg} = E_\varepsilon k\left(1 - \frac{T_l}{T_h}\right) \tag{39}$$

The notation k is used to define the regenerative loss factor:

$$k = \frac{1 - \eta_{reg}}{\ln \varepsilon(\gamma - 1)} \tag{40}$$

Heat quantities change in the case of imperfect regeneration:

$$|Q_h| = |Q_{34}| - Q_{p,reg} = E_\varepsilon\left[1 - k\left(1 - \frac{T_l}{T_h}\right)\right] \tag{41}$$

$$Q_l = Q_{12} - Q_{p,reg} = E_\varepsilon\left[\frac{T_l}{T_h} - k\left(1 - \frac{T_l}{T_h}\right)\right] \tag{42}$$

The mechanical work consumed per cycle in absolute value results in:

$$W = |Q_h| - Q_l \tag{43}$$

It should be mentioned here that the mechanical work consumed in a cycle is independent of the regeneration efficiency η_{reg}.

Using Equation (41), the balance of heat flows at the hot/cold source are obtained:

$$\left|\dot{Q}_h\right| = n|Q_h| = nE_\varepsilon\left[1 - k\left(1 - \frac{T_l}{T_h}\right)\right] = K_h(T_h - T_{wh}) \tag{44}$$

$$\dot{Q}_l = nQ_l = nE_\varepsilon\left[\frac{T_l}{T_h} - k\left(1 - \frac{T_l}{T_h}\right)\right] = K_l(T_{wl} - T_l) \tag{45}$$

The COP performance coefficient of the Stirling refrigeration machine can be determined with the equation:

$$COP = \frac{Q_l}{W} \tag{46}$$

3. Results and Discussions

3.1. Experimental Results

The considered experimental device is a reversible thermal machine (motor and/or receiver) that operates between two heat sources at constant temperature. It works according to the Stirling cycle.

The Stirling refrigerator analyzed is equipped with several sensors: thermocouples, position sensors, pressure sensors, instantaneous position piston sensor, and a device composed of photodiodes and a drilled disk to measure the speed of rotation of the flywheel. The rotation speed n of the electric motor can be varied by means of a control and adjustment device.

The refrigerating power of the analyzed cooling system is estimated by a compensation method by means of a small electric resistance placed inside the cold room (located at the top of the cylinder). In this way, the air at the top of the cylinder cools and heats at the same time. A temperature equal to the ambient temperature can be set inside the cylinder in order to limit the losses through the cylinder wall. In this way, the refrigeration power that corresponds to the heat flow rate taken from the cylinder head of the refrigeration machine is determined with the relation:

$$\dot{Q}_l = UI \tag{47}$$

where U is the voltage (V) and I is the intensity of the electric current (A) corresponding to the electrical compensation resistance.

The thermal conductivity of the cold tank wall can be determined starting from the relation:

$$\dot{Q}_l = K_l \Delta T_l, \tag{48}$$

where

$\dot{Q}_l$ is the refrigerating power of the cooling system, determined by the compensation method, (W); $\Delta T_l = T_l - T_{wl}$, with T_l representing the gas temperature measured inside the cold volume (K); and T_{wl} is the wall temperature of the cold volume, measured with a thermocouple (K).

From the relation of thermal conductivity, we can calculate the global heat exchange coefficient:

$$h = \frac{K_l}{A_l}, \tag{49}$$

where A_l is the contact area of the cold exchanger (upper part of the cylinder) (m^2).

The parameters T_l, T_{wl} and $\dot{Q}_l$ are experimentally determined for several operating modes. The heat transfer coefficient h is calculated for each speed; as expected, and according to the existing data in the literature [30], the overall heat transfer coefficient h increases with increasing rotational speed n (Figure 8).

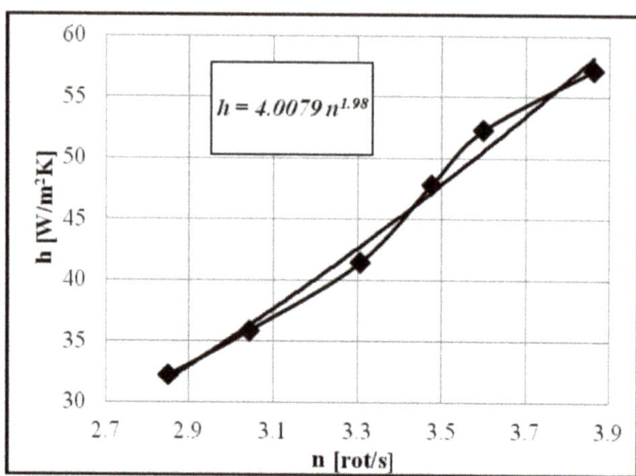

Figure 8. Variation of overall heat transfer coefficient h depending on rotational speed n.

The variation of cooling water temperature and the circulating water flow rate allow us to calculate the power yielded to the water, with the relation:

$$\dot{Q}_h = \dot{m}_w c_w \Delta T_w, \tag{50}$$

where

c_w is water-specific heat (J/kgK); and $\Delta T_w = T_w^e - T_w^i$, with T_w^e representing the output water temperature (K) and T_w^i the input water temperature.

It can be concluded from Table 1 that as the rotation speed n increases, the temperature of the cold gas increases and the temperature of the hot gas decreases. The difference between the two temperatures decreases, which implies increased COP of the refrigerating machine with increased speed.

Table 1. Centralization of experimental data obtained for the Stirling refrigerator.

n (rot/s)	T_l (K)	T_h (K)	ΔT (K)	$\dot{Q}_l$ (W)	$\dot{Q}_h$ (W)	$\dot{W}_{exp}$ (W)	COP_{exp} (–)
2.85	249	348.69	99.69	12.35	32.57	20.22	0.61
3.04	249.5	343.99	94.49	13.40	33.56	20.16	0.66
3.31	250.3	338.03	87.73	14.70	34.54	19.84	0.74
3.47	250.4	332.72	82.32	16.50	35.53	19.03	0.87
3.60	249	330.34	81.34	17.94	37.51	19.57	0.92
3.86	250.7	329.10	78.40	19.20	39.48	20.28	0.95

3.2. Thermodynamic Analysis and Analytical Simulation Results

Using geometric and functional parameters (Table 2) measured or determined by the acquisition program (CassyLab) and using the calculation algorithm, the following developments are obtained depending on the engine rotation speed.

Table 2. Dimensional data of the actual engine.

A_h (m^2)	A_l (m^2)	$V_{min} \cdot 10^{-4}$ (m^3)	$V_{max} \cdot 10^{-4}$ (m^3)	$D_p=D_d$ (m)	$C_p=C_d$ (m)	φ_0 (°)
0.01885	0.03717	1.906	3.278	0.06	0.0484	110

The rotation speed of cooling varied between 2.5 and 3.86 rot/s during the tests, when air was used as working fluid. Any fluid change is not appropriate, as this system has academic use and benefit. The pressure load is 1 bar and should remain, so this refrigerator requires a small amount of mechanical power. The two pistons perform alternating reciprocating motion with an angle of 110°.

The initial data of the simulated point (n = 3.86 rot/s) are listed in Table 3.

Table 3. Initial conditions of a simulated point.

	p_{min} = 70,000 Pa			p_{max} = 211,600 Pa		
n (rot/s)	T_l (K)	T_{wl} (K)	T_h (K)	T_{wh} (K)	T_0 (K)
3.86	250.7	268.5	329.1	295	293

Comparable values were obtained for the exchanged heat flow rates and the mechanical power required to operate the refrigerating machine (Table 4), calculated by processing the experimental data with the two calculation models, 0-D and FPDT.

Table 4. Analyzed heat flow rates.

	Experiment	0-D Model	0-D Error (%)	FPDT Model	FPDT Error (%)
$\dot{Q}_l$(W)	19.21	29.17	51.92	17.79	7.39
$\left|\dot{Q}_h\right|$(W)	39.47	40.22	1.90	37.48	5.04
$\dot{W}_l$(W)	20.28	11.26	44.47	19.68	2.95

The differences obtained between the experimentally processed values and those obtained using of the Schmidt analysis model with imperfect heat regeneration (0-D model) and FPDT model are also reflected in the exergetic calculation of the exchangers. (Tables 5 and 6).

(a) Cold-End Heat Exchanger

Table 5. Exergetic calculation of cold-end heat exchanger.

	Experiment	0-D Model	0-D Error (%)	FPDT Model	FPDT Error (%)
$\left\|\dot{E}x_{Q_l}^{T_l}\right\|$ (W)	3.24	4.92	51.85	3	7.41
$\dot{E}x_{Q_{wl}}^{T_{wl}}$ (W)	1.75	2.66	52	1.62	7.43
$\dot{E}x_l^D$ (W)	1.49	2.26	51.67	1.38	7.38
$\eta_{ex_l}(\%)$	54.01	54.06	0.09	54.08	0.13
$\zeta_l(\%)$	45.98	45.93	0.10	46	0.15

(b) Hot-End Heat Exchanger

Table 6. Exergetic calculation of the hot-end heat exchanger.

	Experiment	0-D Model	0-D Error (%)	FPDT Model	FPDT Error (%)
$\left\|\dot{E}x_{Q_h}^{T_h}\right\|$ (W)	4.33	4.41	1.85	4.11	5.08
$\dot{E}x_{Q_{wh}}^{T_{wh}}$ (W)	0.27	0.27	0	0.25	7.40
$\dot{E}x_h^D$ (W)	4.06	4.13	1.70	3.86	4.92
$\eta_{ex_h}(\%)$	6.23	6.12	1.76	6.18	0.80
$\zeta_h(\%)$	93.76	93.65	0.11	93.91	0.16

A flowchart of the exergy balance equation [31] (Equation (27)) is presented in Figure 9.

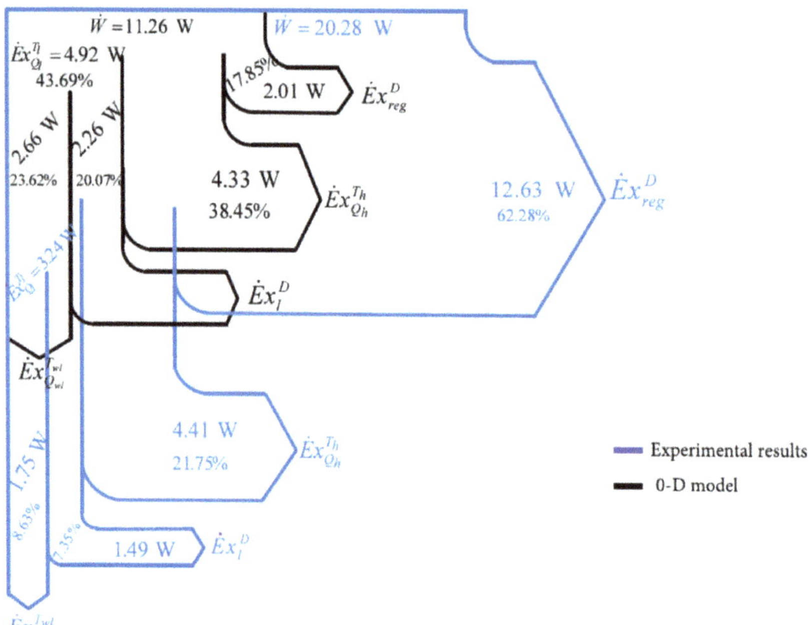

Figure 9. Flowchart of exergy balance equation.

The percentage share in the mechanical power input of each exergy current is presented in Figure 9 as well. Figure 9 shows the difference between the results obtained based on the experimental values presented by comparison with those obtained using the 0-D model. The irreversibility at the level of the regenerator is more important than that at the level of the heat exchangers.

The values obtained when applying the global exergetic efficiency formula $\left(\eta_{EX} = \dfrac{\dot{Ex}_{Q_{wl}}^{T_{wl}}}{\dot{W}} \right)$ are shown in Table 7.

Table 7. Global exergetic efficiency values.

	Experiment	0-D Model	FPDT Model
$\eta_{EX}(\%)$	8.63	23.65	8.23

Table 8 compares the results of the two models of thermodynamic analysis and the experimentally obtained results for the Stirling machine.

Table 8. Comparison of experimental and analytical results from analyzed methods.

n = 3.86 (rot/min)					
COP_{exp}	$\dot{W}_{exp}$	COP_{0-D}	$\dot{W}_{0-D}$	COP_{FPDT}	$\dot{W}_{FPDT}$
(–)	(W)	(–)	(W)	(–)	(W)
0.947	20.280	2.570	11.260	0.905	19.68

The experimental COP of the refrigerating machine for a cold temperature of $-22.45\,^{\circ}\text{C}$ is found to be 0.947. From the same cold temperature, the COP obtained after applying the 0-D model was found to be 2.57, with an error of 171.38%.

Applying the FPDT model returns a value of 0.905 for the COP. This shows that the simulation results approach the experimental results with an error of 4.43%.

In terms of evaluating the mechanical power input at the same speed, n = 3.86 rot/min (corresponding to the cold space air temperature $t_h = -22.45\,^{\circ}\text{C}$), after numerical simulation the 0-D model returns a value of 11.26 W (error of 44.47%), while in the FPDT model, the mechanical power required for operating the Stirling refrigerator is 19.68 W, with an error of 2.95%.

In addition, and after comparing the values of the global exergetic efficiency (Table 7) obtained for the two proposed thermodynamic analysis methods, the 0-D model provides an η_{EX} of 23.65 with an error of 174%, while the global exergetic efficiency calculated with the FPDT model is 8.23 (error of 4.63%).

4. Conclusions

A 0-D numerical model describing the evolution of variables (pressure, volume, mass, exchanged energy, irreversibility) as a function of the crankshaft angle is presented. The model uses the energy and exergy balance in a controlled volume, assuming a steady-state operation in the Stirling refrigerator, in order to obtain the overall irreversibility of the heat exchangers. External irreversibility is due to a finite temperature difference between gas and heat exchangers, while internal irreversibility is due to regenerative heat loss and entropy generation. It is found that the irreversibility at the level of the regenerator is more important than that at the level of the heat exchangers (Figure 9).

A flowchart of the exergy balance of the Stirling refrigerator is presented to show the internal and external irreversibilities (destroyed exergy flow). In the flow diagram (Figure 9), the exergy flows of the working gas with two reservoirs (heat from hot source and heat to cold sink) are shown at different temperatures, T_h and T_{wh} for the source and T_l and T_{wl} for the sink.

The study was completed by comparing the results obtained with the isothermal model and the FPDT model. The irreversibilities that FPDT model takes into account are exo-irreversibilities due to the finite heat transfer between the sources (hot source, cold source, regenerator) and the working fluid.

Regarding the evaluation of the mechanical power necessary for operating the refrigeration machine using the Schmidt isothermal model with imperfect regeneration, the difference between the experimental results and the results given by the thermodynamic model is justified by the fact that friction and aerodynamic losses are not taken into account in this model.

The results of the two thermodynamic models are presented in comparison with the experimental results, which leads to validation of the proposed FPDT model for the functional and constructive parameters of the studied refrigerating machine. It is found that the calculated values are very close to the experimental values, which validates the proposed analysis model for the β-type Stirling refrigerator. Therefore, the FPDT model proves to be a useful tool for analyzing the performance (COP and input power) of β-type Stirling refrigeration machines.

Author Contributions: Conceptualization, C.D. and L.G.; Methodology, C.D. and L.G.; Software, C.D.; Validation, C.D. and A.D.; Formal analysis, C.D., L.G. and A.D.; Investigation, C.D.; Resources, C.D. and L.G.; Data curation, C.D., G.C. and M.C.; Writing—original draft preparation, C.D., M.C. and G.C.; writing—review and editing, C.D., G.C. and M.C.; Visualization, C.D., M.C. and G.C.; supervision, A.D.; Project administration, C.D.; Funding acquisition, G.C. All authors have read and agreed to the published version of the manuscript.

Funding: The work has been funded by the Romanian Ministry of Education and Politehnica University of Bucharest through the PubArt programme.

Institutional Review Board Statement: Not applicable.

Informed Consent Statement: Not applicable.

Data Availability Statement: The data presented in this study are available on request from the corresponding author.

Conflicts of Interest: The authors declare no conflict of interest.

Abbreviations

A	heat exchange surface, m^2
c	specific heat, Jkg^{-1}K^{-1}
C	stroke of the piston, m
D	diameter of the piston, m
Ex	Exergy, J
$\dot{E}_x$	exergy flow rate, W
h	convective heat transfer coefficient, Wm^{-2}K^{-1}
I	current, A
k	losses factor in regenerator,
K	heat exchanger conductance, WK^{-1}
m	mass, kg
n	engine rotation speed, rot·s^{-1}
p	pressure, Pa
Q	heat, J
$\dot{Q}$	heat transfer rate, W
R	gas constant, Jkg^{-1}K^{-1}
S	entropy, JK^{-1}
s	specific entropy, Jkg^{-1}K^{-1}
T	temperature, K
U	voltage, V

V	volume, m^3
W	work, J
$\dot{W}$	mechanical power, W
Greek symbols	
γ	adiabatic exponent, -
φ	rotation angle, °
φ_0	phase lag angle, °
ε	volumetric compression ratio (V_{max}/V_{min}), -
η	efficiency, -
Π	entropy increase, JK^{-1}
$\dot{\Pi}$	rate of entropy increase, WK^{-1}
ζ	dissipation coefficient, -
Subscripts	
C	compression
ε	depending on ε
ex	exergetic
E	expansion
d	displacer
h	hot on working gas side
l	low on working gas side
m	dead
max	maximum
min	minimum
p	piston
rev	reversible
reg	regenerator
v	constant volume (specific heat)
wl	wall on source side
wh	wall on sink side

References

1. Getie, M.Z.; Lanzetta, F.; Bégot, S.; Admassu, B.T.; Hassen, A.A. Reversed regenerative Stirling cycle machine for refrigeration application: A review. *Int. J. Refrig.* **2020**, *118*, 173–187. [CrossRef]
2. Jacob, W.L.K. The Stirling refrigeration cycle in cryogenic technology. *Adv. Sci.* **1968**, *25*, 261.
3. Beale, W. Stirling Cycle Type Thermal Device. US3552120A, 5 January 1971.
4. Chen, J.; Yan, Z. Regenerative characteristics of magnetic or gas Stirling refrigeration cycle. *Cryogenics* **1993**, *33*, 863–867. [CrossRef]
5. Toghyani, S.; Kasaeian, A.; Hashemabadi, S.H.; Salimi, M. Multi-objective optimization of GPU3 Stirling engine using third order analysis. *Energy Convers. Manag.* **2014**, *87*, 521–529. [CrossRef]
6. Curzon, F.L.; Ahlborn, B. Efficiency of a Carnot engine at maximum power output. *Am. J. Phys.* **1975**, *43*, 22–24. [CrossRef]
7. Chen, L.; Wu, C.; Sun, F. Cooling load versus cop characteristics for an irreversible air refrigeration cycle. *Energy Convers. Manag.* **1998**, *39*, 117–125. [CrossRef]
8. Martaj, N.; Grosu, L.; Rochelle, P. Exergetical analysis and design optimization of the Stirling engine. *Int. J. Exergy* **2006**, *3*, 45–67. [CrossRef]
9. Feidt, M. Optimal use of energy systems and processes. *Int. J. Energy* **2008**, *5*, 500–531. [CrossRef]
10. Petrescu, S.; Costea, M.; Harman, C.; Florea, T. Application of the Direct Method to Irreversible Stirling cycles with finite speed. *Int. J. Energy Res.* **2002**, *26*, 589–609. [CrossRef]
11. Grosu, L.; Dobre, C.; Petrescu, S. Study of a Stirling engine used for domestic micro-cogeneration. Thermodynamic analysis and experiment. *Int. J. Energy Res.* **2015**, *39*, 1280–1294. [CrossRef]
12. Chen, J. Minimum power input of irreversible Stirling refrigerator for given cooling rate. *Energy Convers. Manag.* **1998**, *39*, 1255–1263. [CrossRef]
13. Otaka, T.; Ota, M.; Murakami, K.; Sakamoto, M. Study of performance characteristics of a small Stirling refrigerator. *Heat Transf. Asian Res.* **2002**, *31*, 344–361. [CrossRef]
14. Ataer, Ö.E.; Karabulut, H. Thermodynamic analysis of the v-type Stirling-cycle refrigerator. *Int. J. Refrig.* **2005**, *28*, 183–189. [CrossRef]
15. McFarlane, P.K. Mathematical Model and Experimental Design of an Air-filled Alpha Stirling Refrigerator. Bachelor's Thesis, University of Notre Dame, Notre Dame, Indiana, 2014.
16. Li, R.; Grosu, L. Parameter effect analysis for a Stirling cryocooler. *Int. J. Refrig.* **2017**, *80*, 92–105. [CrossRef]

17. Hachem, H.; Gheith, R.; Nasrallah, S.B.; Aloui, F. Impact of operating parameters on beta type regenerative Stirling machine performances. In Proceedings of the ASME/JSME/KSME 2015 Joint Fluids Engineering Conference, Seoul, Korea, 26–31 July 2015; American Society of Mechanical Engineers: New York, NY, USA, 2015.
18. Hachem, H.; Gheith, R.; Aloui, F.; Nasrallah, S.B. Optimization of an air-filled beta type Stirling refrigerator. *Int. J. Refrig.* **2017**, *76*, 296–312. [CrossRef]
19. Dai, D.D.; Yuan, F.; Long, R.; Liu, Z.C.; Liu, W. Imperfect regeneration analysis of Stirling engine caused by temperature differences in regenerator. *Energy Convers. Manag.* **2018**, *158*, 60–69. [CrossRef]
20. Dai, D.D.; Yuan, F.; Long, R.; Liu, Z.C.; Liu, W. Performance analysis and multi-objective optimization of a Stirling engine based on MOPSOCD. *Int. J. Sci.* **2018**, *124*, 399–406. [CrossRef]
21. Petrescu, S.; Florea, T.; Harman, C.; Costea, M. A method for calculating the coefficient for regenerative losses in Stirling machines. In Proceedings of the 5th European Stirling Formum, Osnabrick, Germany, 22–24 February 2000; pp. 121–129.
22. Kongtragool, B.; Wongwises, S. Thermodynamic analysis of a Stirling engine including dead volumes of hot space, cold space and regenerator. *Renew. Energy* **2006**, *3*, 345–359. [CrossRef]
23. Popescu, G.; Radcenco, V.; Costea, M.; Feidt, M. Finite-time thermodynamics optimisation of an endo-exo-irreversible Stirling engine. *Rev. Gen. Therm.* **1996**, *35*, 656–661. [CrossRef]
24. Grosu, L.; Rochelle, P.; Martaj, N. Thermodynamique a échelle finie: Optimisation du cycle moteur de Stirling pour l'ingénieur. *Int. J. Exergy* **2006**, *3*, 1.
25. Li, R.; Grosu, L.; Queiro-Conde, D. Multy-objective optimization of Stirling engine using Finite Physical Dimensions Thermodynamics (FPDT) Method. *Energy Convers. Manag.* **2016**, *124*, 517–527. [CrossRef]
26. Li, R.; Grosu, L.; Queiros-Condé, D. Losses effect on the performance of a Gamma type Stirling engine. *Energy Convers. Manag.* **2016**, *114*, 28–37. [CrossRef]
27. Martaj, N.; Bennacer, R.; Grosu, L.; Savarese, S.; Laaouatni, A. LTD Stirling engine with regenerator. Numerical and experimental study. *Mech. Ind.* **2017**, *18*, 305. [CrossRef]
28. Dobre, C.; Grosu, L.; Costea, M.; Constantin, M. Beta Type Stirling Engine. Schmidt and Finite Physical Dimensions Thermodynamics Methods Faced to Experiments. *Entropy* **2020**, *22*, 1278. [CrossRef] [PubMed]
29. Rochelle, P.; Grosu, L. Analytical solutions and optimization of the exoirreversible Schmidt cycle with imperfect regeneration for the 3 classical types of Stirling engine. *Oil Gas Sci. Technol.* **2011**, *66*, 747–758. [CrossRef]
30. Bădescu, V.; Popescu, G.; Feidt, M. Model of optimized solar heat engine operating on Mars. *Energy Convers. Manag.* **1999**, *40*, 813–819. [CrossRef]
31. Dincer, I.; Rosen, M.A. *Exergy, Energy, Environment and Sustainable Development*; Elsevier: Amsterdam, The Netherlands, 2020. ISBN 9780128243930.

 *entropy*

Article

Power and Efficiency Optimization for Open Combined Regenerative Brayton and Inverse Brayton Cycles with Regeneration before the Inverse Cycle

Lingen Chen [1,2,*], Huijun Feng [1,2] and Yanlin Ge [1,2]

[1] Institute of Thermal Science and Power Engineering, Wuhan Institute of Technology, Wuhan 430205, China; huijunfeng@139.com (H.F.); geyali9@hotmail.com (Y.G.)

[2] School of Mechanical & Electrical Engineering, Wuhan Institute of Technology, Wuhan 430205, China

* Correspondence: lingenchen@hotmail.com

Received: 29 May 2020; Accepted: 16 June 2020; Published: 17 June 2020

Abstract: A theoretical model of an open combined cycle is researched in this paper. In this combined cycle, an inverse Brayton cycle is introduced into regenerative Brayton cycle by resorting to finite-time thermodynamics. The constraints of flow pressure drop and plant size are taken into account. Thirteen kinds of flow resistances in the cycle are calculated. On the one hand, four isentropic efficiencies are used to evaluate the friction losses in the blades and vanes. On the other hand, nine kinds of flow resistances are caused by the cross-section variances of flowing channels, which exist at the entrance of top cycle compressor (TCC), the entrance and exit of regenerator, the entrance and exit of combustion chamber, the exit of top cycle turbine, the exit of bottom cycle turbine, the entrance of heat exchanger, as well as the entrance of bottom cycle compressor (BCC). To analyze the thermodynamic indexes of power output, efficiency along with other coefficients, the analytical formulae of these indexes related to thirteen kinds of pressure drop losses are yielded. The thermodynamic performances are optimized by varying the cycle parameters. The numerical results reveal that the power output presents a maximal value when the air flow rate and entrance pressure of BCC change. In addition, the power output gets its double maximal value when the pressure ratio of TCC further changes. In the premise of constant flow rate of working fuel and invariant power plant size, the thermodynamic indexes can be optimized further when the flow areas of the components change. The effect of regenerator on thermal efficiency is further analyzed in detail. It is reported that better thermal efficiency can be procured by introducing the regenerator into the combined cycle in contrast with the counterpart without the regenerator as the cycle parameters change in the critical ranges.

Keywords: combined cycle; inverse Brayton cycle; regenerative Brayton cycle; power output; thermal efficiency; finite time thermodynamics

---

1. Introduction

A theoretical model of an open combined Brayton cycle (OCBC) was built by Chen et al. [1] on the bases of the models provided by Refs. [2–15]. In the OCBC model built in Ref. [1], an inverse Brayton cycle was introduced into regenerative Brayton cycle by resorting to the finite-time thermodynamics (FTT) [16–30], which has been applied for various processes and cycles [31–40]. The thermodynamic indexes of the OCBC have been analyzed in Ref. [1]. In order to further optimize the thermodynamic indexes, such as the power output (PO), thermal efficiency (TE), and pressure ratio (PR) of top cycle compressor (TCC), the analytical formulae related with 13 kinds of pressure drop losses (PDLs) are yielded. These PDLs take place in the whole cycle, such as the combustion chamber, the compressors, the regenerator, the turbines, as well as various flow processes. By employing the similar principle according to Refs. [41–47] and the method according to Refs. [2–7,12–15], the PO and TE will be numerically optimized in this paper.

In this paper, the performance optimizations of the OCBC will be conducted by means of varying the PR of TCC, mass flow rate (MFR), as well as PDL allocation. The maximum PO and TE of the OCBC will be gained after optimizations. Furthermore, the influences of cycle parameters on the optimal results will be numerically yielded.

2. Brief Introduction of the OCBC Model

Alabdoadaim et al. [11] proposed new configuration of an OCBC. It has a top cycle and a bottom cycle. The former is a regenerative Brayton cycle and is applied as a gas generator to power bottom one. The latter is an inverse Brayton cycle. The PO of the OCBC is totally produced by bottom cycle. As shown in Figure 1 [1,11], the top cycle contains compressor 1 (top cycle compressor (TCC)), regenerator, combustion chamber, and turbine 1 (top cycle turbine), whereas the bottom cycle contains turbine 2 (bottom cycle turbine), heat exchanger, and compressor 2 (bottom cycle compressor (BCC)).

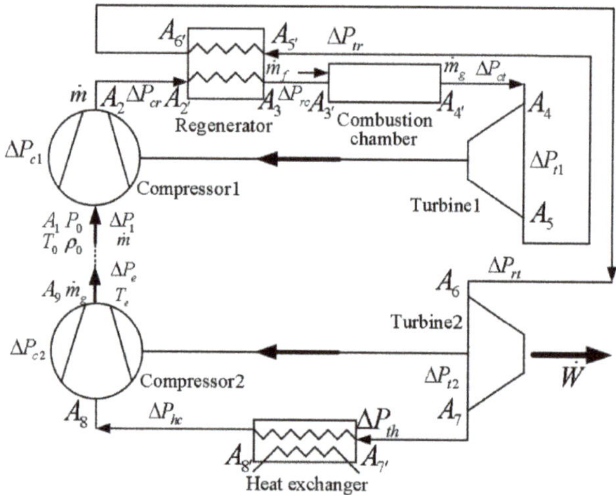

Figure 1. Pressure drop loss (PDL) and mass flow rate (MFR) distributions for the combined regenerative Brayton and inverse Brayton cycles [1,11].

According to FTT theory for open cycles [2–7,12–15], there are 13 kinds of flow resistances in the OCBC, 4 of them are evaluated by isentropic efficiencies of turbines and compressors, which take into account the friction losses in the blades and vanes, and the other nine kinds of them are caused by the cross-section variances of flowing channels, which exist at the entrance of TCC, the entrance and exit of regenerator, the entrance and exit of combustion chamber, the exit of turbine 1, the exit of turbine 2, the entrance of heat exchanger, as well as the entrance of BCC.

The model of the OCBC, which is expressed using PDL and MFR distributions and temperature–entropy diagram, is shown in Figure 2 [1].

According to Chen et al. [1], after analyzing the OCBC, all of the PDLs in the system can be expressed as functions of the relative PD (ψ_1) of the entrance of TCC, $\psi_1 = \Delta P_1/P_0$, where P_0 is the atmosphere pressure and $\Delta P_1 = K_1(\rho_0 V_1^2/2)$ is the PD of the entrance of TCC, where K_1 is the contraction pressure loss coefficient and V_1 is average air velocity through the entrance flow cross-section A_1 (see Figure 1).

Besides, all of the dimensionless power inputs of the compressors, power outputs of the turbines, as well as the heat transfer rate produced by fuel were obtained [1]; they are functions of the relative PD (ψ_1) of the entrance of TCC:

$$\overline{W}_{c1} = \frac{\gamma_{a1}(\theta_{c1s}-1)}{\eta_{c1}(\gamma_{a1}-1)}\psi_1^{1/2} \tag{1}$$

$$\overline{W}_{c2} = [1+1/(\lambda L_0)]\frac{\tau(T_6/T_5)(\theta_{c2s}-1)\gamma_{gc2}}{\eta_{c2}(\gamma_{gc2}-1)\theta_i\theta_{t1}\theta_{t2}}\psi_1^{1/2} \tag{2}$$

$$\overline{W}_{t1} = [1+1/(\lambda L_0)]\frac{\eta_{t1}\tau(1-1/\theta_{t1s})\gamma_{g1}}{\gamma_{g1}-1}\psi_1^{1/2} \tag{3}$$

$$\overline{W}_{t2} = [1+1/(\lambda L_0)]\frac{\eta_{t2}\tau(1-1/\theta_{t2s})(T_6/T_5)\gamma_{g2}}{(\gamma_{g2}-1)\theta_{t1}}\psi_1^{1/2} \tag{4}$$

$$\overline{Q}_f = \left(1+\frac{1}{\lambda L_0}\right)\frac{\gamma_{gc}(\tau-T_3/T_0)}{(\gamma_{gc}-1)\eta_{cf}}\psi_1^{1/2} \tag{5}$$

where γ_{a1} is air specific heat ratio, $\theta_{c1s}=T_{2s}/T_1=\beta_{c1}^{(\gamma_{a1}-1)/\gamma_{a1}}$ is isentropic temperature ratio of TCC, $\beta_{c1}=P_2/P_1=\beta_1/(1-\psi_1)$ is effective pressure ratio (PR) of TCC, $\beta_1=P_2/P_0$ is apparent compressor PR, η_{c1} is isentropic efficiency of TCC; L_0 and λ are theoretical air quantity and excess air ratio of the combustor, $\tau=T_4/T_0$, γ_{gc2} is gas specific heat ratio in turbine 2, $\theta_{c2s}=T_{9s}/T_8=\beta_{c2}^{(\gamma_{gc2}-1)/\gamma_{gc2}}$ is isentropic temperature ratio of turbine 2, $\beta_{c2}=P_9/P_8$, η_{c2} is isentropic efficiency of BCC, $\theta_i=T_7/T_8$, $\theta_{t2}=T_{6'}/T_7$, $\theta_{t1}=T_4/T_5$; η_{t1} is isentropic efficiency of turbine 1, γ_{g1} is gas specific heat ratio in turbine 1, $\theta_{t1s}=T_4/T_{5s}=\beta_{t1}^{(\gamma_{g1}-1)/\gamma_{g1}}$ is isentropic temperature ratio of turbine 1, $\beta_{t1}=P_4/P_5$; η_{t2} is isentropic efficiency of turbine 2, γ_{g2} is gas specific heat ratio in turbine 2, $\theta_{t2s}=T_{6'}/T_{7s}=\beta_{t2}^{(\gamma_{g2}-1)/\gamma_{g2}}$ is isentropic temperature ratio of turbine 2, $\beta_{t2}=P_6/P_7$; γ_{gc} is specific heat ratio in combustor; and η_{cf} is combustor efficiency. All of the specific heat ratios for air and gas are evaluated according to empirical correlation based on averaged temperatures of air and gas [48,49].

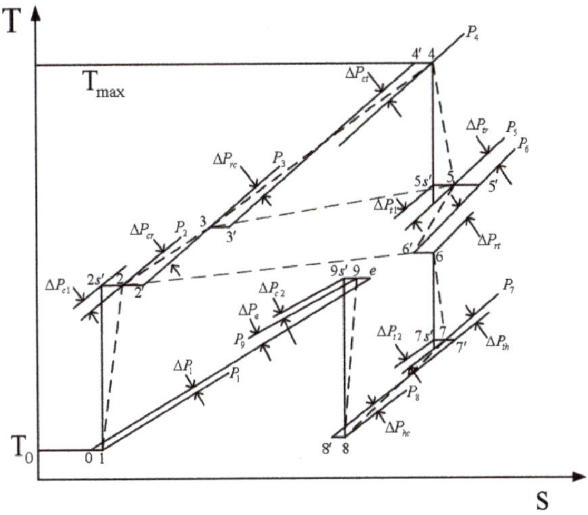

Figure 2. Temperature–entropy diagram and the flow resistances for the combined regenerative Brayton and inverse Brayton cycles [1].

According to two operation principle [11], one has $\overline{W}_{c1} = \overline{W}_{t1}$. The net PO and TE are as follows [1]:

$$\dot{W} = \dot{W}_{t2} - \dot{W}_{c2} = [1 + 1/(\lambda L_0)]\{\frac{\eta_{t2}(1 - 1/\theta_{t2s})(T_6/T_5)\gamma_{g2}}{(\gamma_{g2} - 1)\theta_{t1}} - \frac{(T_6/T_5)(\theta_{c2s} - 1)\gamma_{gc2}}{\eta_{c2}(\gamma_{gc2} - 1)\theta_i\theta_{t1}\theta_{t2}}\}\tau\psi_1^{1/2} \quad (6)$$

$$\eta_1 = \frac{(\gamma_{gc} - 1)\eta_{cf}\tau}{\gamma_{gc}(\tau - T_3/T_0)}\left[\frac{\eta_{t2}(1 - 1/\theta_{t2s})(T_6/T_5)\gamma_{g2}}{(\gamma_{g2} - 1)\theta_{t1}} - \frac{(T_6/T_5)(\theta_{c2s} - 1)\gamma_{gc2}}{\eta_{c2}(\gamma_{gc2} - 1)\theta_i\theta_{t1}\theta_{t2}}\right] \quad (7)$$

3. Power Output Optimization

In this section, a series of numerical solutions are conducted to examine the influences of PR of bottom cycle, MFR of working air, as well as PDs on the net PO. In order to carry out numerical examples, the pertinent variation ranges and values of the cycle parameters are listed as: $0 \leq \psi_1 \leq 0.2$, $5 \leq \beta_1 \leq 40$, $1 \leq \beta_i \leq 2.5$, $4 \leq \tau \leq 6$, $P_0 = 0.1013MPa$, $T_0 = 300K$, $\eta_{c1} = 0.9$, $\eta_{c2} = 0.87$, $\eta_{t1} = 0.85$, $\eta_{t2} = 0.83$, $\eta_{cf} = 0.99$, $\varepsilon = 0.9$, and $\varepsilon_R = 0.9$ [2,3,11]. In addition, ratio of the outermost equivalent flow cross-sections (entrance of TCC/outlet of BCC) covered the range $0.25 \leq a_{1-9} \leq 4$, where a_{1-9} is the dimensionless group [1–3]:

$$a_{1-9} = \frac{A_1}{A_9}\left(\frac{K_9}{K_1}\right)^{1/2} \quad (8)$$

$$a_{1-i} = \frac{A_1}{A_i}\left(\frac{K_i}{K_1}\right)^{1/2}, i = 2,3,4,5,6,7,8,9 \quad (9)$$

where $a_{1-2} = a_{1-3} = a_{1-5} = a_{1-6} = a_{1-7} = a_{1-8} = a_{1-9} = 1/3$, $a_{1-4} = 1/2$, and $0.25 \leq a_{1-9} \leq 4$ are selected [1–3].

Figures 3–6 present the relationships of the maximum dimensionless PO ($\overline{W}_{max}$) of the OCBC, relative optimal PR ($(\beta_{iopt})_W$) of BCC, as well as optimal entrance PD ($(\psi_{1opt})_W$) of TCC versus the PR (β_1) of TCC, temperature ratio (τ) of top cycle (TC), effectiveness (ε) of heat exchanger, as well as the effectiveness (ε_R) of regenerator, respectively. On the one hand, it is manifest that $\overline{W}_{max}$ exhibits an increasing trend as τ and ε increase. However, it exhibits a decreasing trend as ε_R increases. $\overline{W}$ can be twice maximized ($\overline{W}_{max,2}$) at the $(\beta_{1opt})_W$. On the other hand, it can also be found that $(\beta_{iopt})_W$ increases as β_1 and ε increase, while it decreases as τ and ε_R increase. It is obvious that the relationships of $(\psi_{1opt})_W$ versus β_1 and ε_R exhibit the parabolic-like curves. $(\psi_{1opt})_W$ increases as τ increases because the larger τ corresponds lager MFR of the working air.

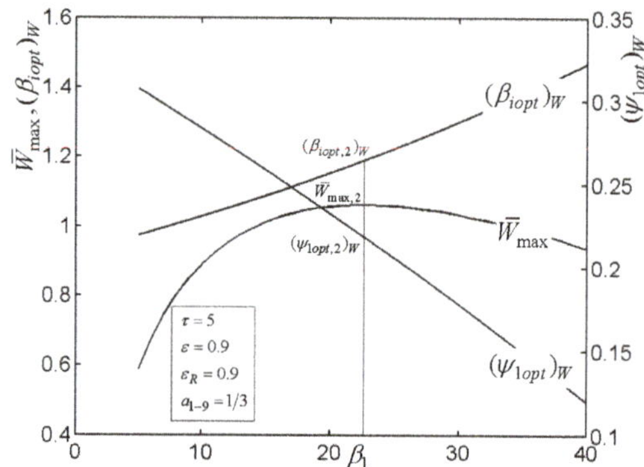

Figure 3. Relationships of $\overline{W}_{max} - \beta_1$, $(\beta_{iopt})_W - \beta_1$, and $(\psi_{1opt})_W - \beta_1$.

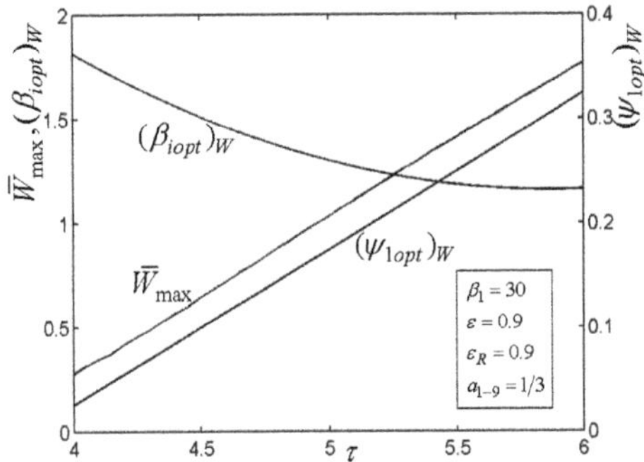

Figure 4. Relationships of $\overline{W}_{max} - \tau$, $(\beta_{iopt})_W - \tau$, and $(\psi_{1opt})_W - \tau$.

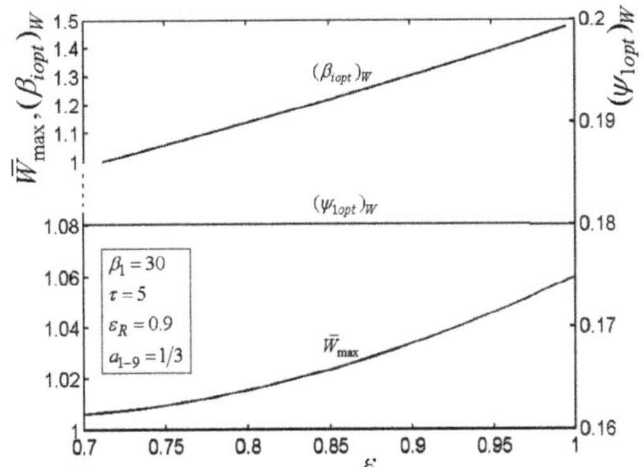

Figure 5. Relationships of $\overline{W}_{max} - \varepsilon$, $(\beta_{iopt})_W - \varepsilon$, and $(\psi_{1opt})_W - \varepsilon$.

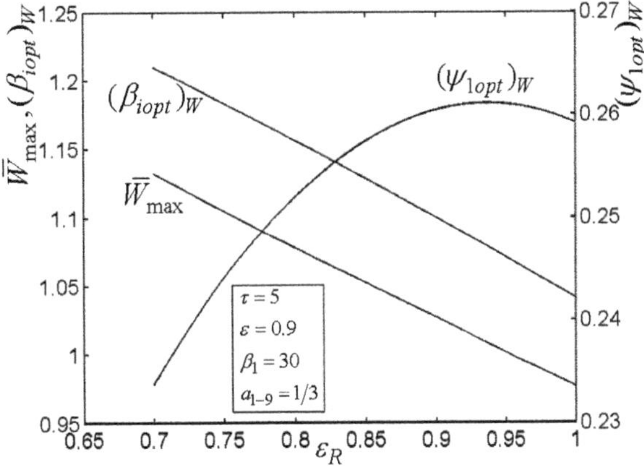

Figure 6. Relationships of $\overline{W}_{max} - \varepsilon_R$, $(\beta_{iopt})_W - \varepsilon_R$, and $(\psi_{1opt})_W - \varepsilon_R$.

Figures 7–12 present the influences of a_{1-9} on the relationships of $\overline{W}_{max,2}$, relative optimal PD $((\psi_{1opt,2})_W)$, $(\beta_{1opt})_W$, as well as relative entrance pressure $((P_{8opt,2})_W)$ of BC versus τ of TC, effectiveness (ε) of heat exchanger, as well as effectiveness (ε_R) of regenerator, respectively. According to these figures for the fixed τ, ε, and ε_R, both $\overline{W}_{max,2}$ and $(\psi_{1opt,2})_W$ decrease as a_{1-9} increases, and on the contrary, both $(\beta_{1opt})_W$ and $(P_{8opt,2})_W$ exhibit an increasing trend as a_{1-9} increases. The twice maximized PO

$(\overline{W}_{\max,2})$ increases about 100% when a_{1-9} decreases from 3 to 1 for the fixed τ. It shows that the size parameters of the entrance of TCC and outlet of BCC affect the performance of OCBC greatly. One can also see that $\overline{W}_{\max,2}$ exhibits an increasing trend as τ and ε increase, while it exhibits a decreasing trend as ε_R increases. It shows that the regeneration cannot increase the PO in the discussed conditions because of the increase of PDL by adding a regenerator. In the case of $a_{1-9} = 1/3$, $(\psi_{1opt,2})_W$ increases as τ and ε increase. In addition, $(\beta_{1opt})_W$ tends to gradually increase as τ and ε_R increase. Besides, $(P_{8opt,2})_W$ will be equal to environment pressure when a_{1-9} is big enough. In this case, the BCC can be disregarded.

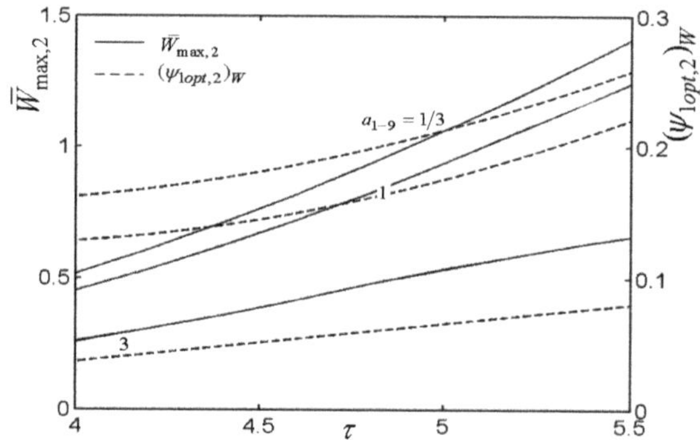

Figure 7. Influences of a_{1-9} on the relationships of $\overline{W}_{\max,2} - \tau$ and $(\psi_{1opt,2})_W - \tau$.

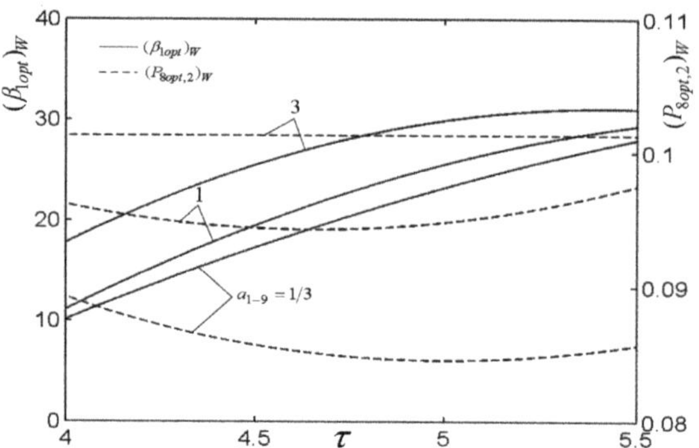

Figure 8. Influences of a_{1-9} on the relationships of $(\beta_{1opt})_W - \tau$ and $(P_{8opt,2})_W - \tau$.

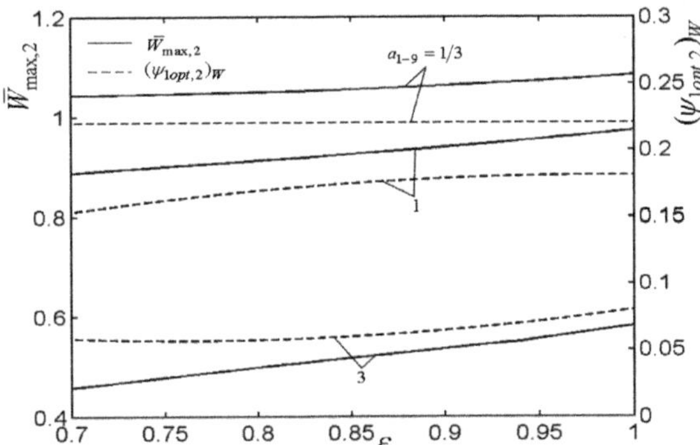

Figure 9. Influences of a_{1-9} on the relationships of $\overline{W}_{max,2} - \varepsilon$ and $(\psi_{1opt,2})_W - \varepsilon$.

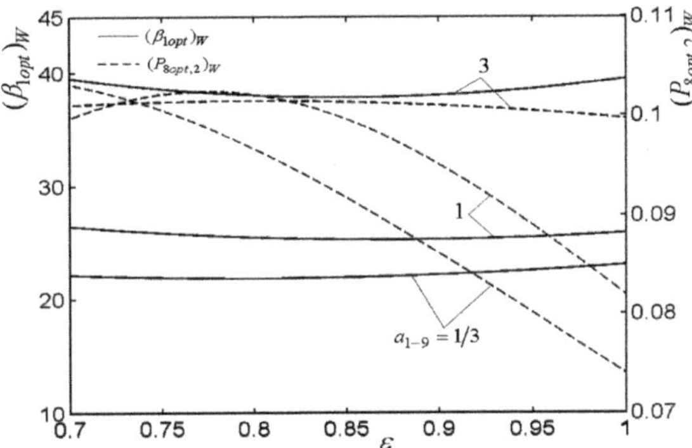

Figure 10. Influences of a_{1-9} on the relationships of $(\beta_{1opt})_W - \varepsilon$ and $(P_{8opt,2})_W - \varepsilon$.

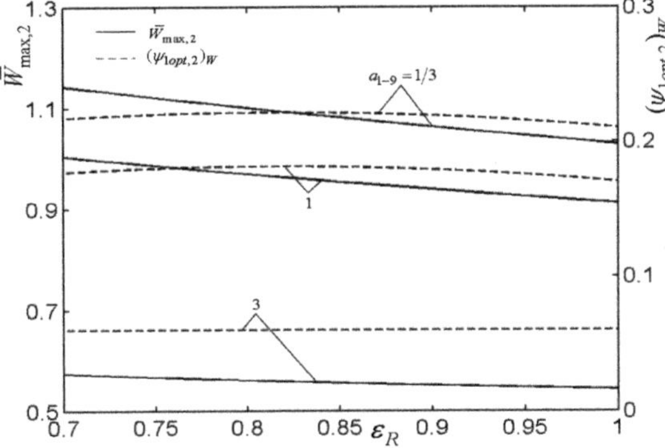

Figure 11. Influences of a_{1-9} on the relationships of $\overline{W}_{max,2} - \varepsilon_R$ and $(\psi_{1opt,2})_W - \varepsilon_R$.

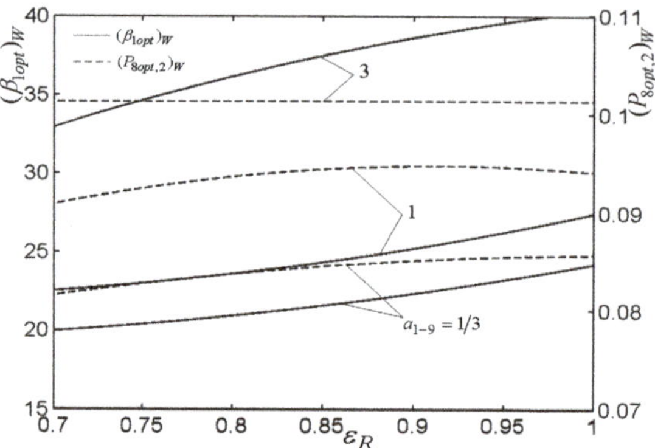

Figure 12. Influences of a_{1-9} on the relationships of $(\beta_{1opt})_W - \varepsilon_R$ and $(P_{8opt,2})_W - \varepsilon_R$.

4. Thermal Efficiency Optimization

In this section, the aforementioned theoretical model is optimized herein by considering two practical constraints. The heat transfer rate $(\dot{Q}_f)$ discharged by working fuel is invariant. As a result, $\dot{Q}_f$ constraint is expressed by [1–3]

$$\dot{Q}_f = A_1(2/K_1)^{1/2}P_0(RT_0)^{1/2}Q_f\psi_1^{1/2}/(\lambda L_0 RT_0) = const \tag{10}$$

In addition, the other constraint is the total size of the OCBC, which is characterized by $A_1 + A_5 + A_7 + A_9$. For simplification, the following constraint considering the areas (A_1 and A_7) of turbine 2 exit and TCC entrance is introduced [1–3]

$$A_1/K_1^{1/2} + A_7/K_7^{1/2} = A_* = const \tag{11}$$

It is used to search for the optimal allocation ratio (x) of flow area defined by $A_1/K_1^{1/2} = xA_*$ and $A_7/K_7^{1/2} = (1-x)A_*$. From Equations (10) and (11), $\overline{Q}_{f*}$ is given as

$$\overline{Q}_{f*} = \dot{Q}_f/\left[A_*P_0(RT_0)^{1/2}\right] = Cx\psi_1^{1/2}/\lambda = const \tag{12}$$

where $C = 2^{1/2}Q_f/(L_0 RT_0)$.

On this basis, the POs of turbine 2 and BCC can be, respectively, calculated as

$$\overline{W}_{t2*} = \frac{\dot{W}_{t2}}{A_*P_0(RT_0)^{1/2}} = [1+1/(\lambda L_0)]\frac{\sqrt{2}x\eta_{t2}\tau(1-1/\theta_{t2s})(T_6/T_5)\gamma_{g2}}{(\gamma_{g2}-1)\theta_{t1}}\psi_1^{1/2} \tag{13}$$

$$\overline{W}_{c2*} = \frac{\dot{W}_{c2}}{A_*P_0(RT_0)^{1/2}} = [1+1/(\lambda L_0)]\frac{\sqrt{2}x\tau(T_6/T_5)(\theta_{c2s}-1)\gamma_{gc2}}{\eta_{c2}(\gamma_{gc2}-1)\theta_i\theta_{t1}\theta_{t2}}\psi_1^{1/2} \tag{14}$$

From Equations (12)–(14), the TE derived by the first law of thermodynamics is written as

$$\eta_1 = \frac{\overline{W}_{t2*} - \overline{W}_{c2*}}{\overline{Q}_{f*}} = \frac{\sqrt{2}\lambda}{C}\left[1 + \frac{1}{\lambda L_0}\right]\left[\frac{\eta_{t2}\tau(1-1/\theta_{t2s})(T_6/T_5)\gamma_{g2}}{(\gamma_{g2}-1)\theta_{t1}} - \frac{\tau(T_6/T_5)(\theta_{c2s}-1)\gamma_{gc2}}{\eta_{c2}(\gamma_{gc2}-1)\theta_i\theta_{t1}\theta_{t2}}\right] \tag{15}$$

Figure 13 presents the relationship of the excessive air ratio (λ) versus relative PD (ψ_1) of TCC entrance. As shown in Figure 13, it is indicated that λ increases as ψ_1 increases. Figure 14 presents the influences of regenerator effectiveness (ε_R) on the relationships of TE (η_1) versus PR (β_i) of BCC, relative

PD (ψ_1) of TCC entrance, as well as area allocation ratio (x). As shown in Figure 14, it is indicated that η_1 can be maximized by selecting optimal values $((\beta_{iopt})_\eta, (\psi_{1opt})_\eta$ and $x_{opt})$ of β_i, ψ_1, and x in the both cases ($\varepsilon_R = 0.9$ and $\varepsilon_R = 0$). Moreover, in the discussed ranges of β_i, ψ_1, and x, the OCBC with regenerator can procure a better TE in contrast with the counterpart without regenerator. It shows that the regeneration can increase the TE.

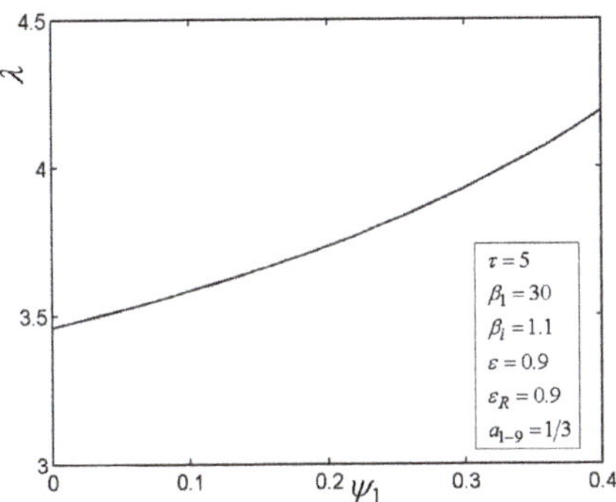

Figure 13. Relationships of $\lambda - \psi_1$.

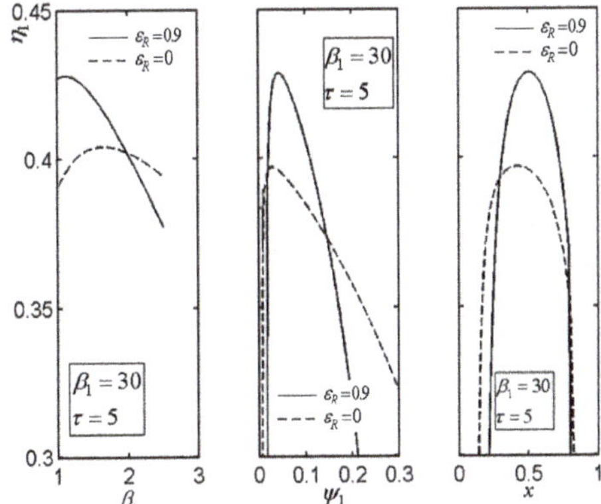

Figure 14. Influences of ε_R on the relationships of $\eta_1 - \beta_i$, $\eta_1 - \psi_1$, and $\eta_1 - x$.

Figure 15, Figure 16, Figure 17, Figure 18 andFigure 19 present the relationships of the maximum TE (η_{1max}), optimal PD $(\psi_{1opt})_\eta$ of TCC entrance, optimal pressure $(P_{8opt})_\eta$ of BCC entrance, as well as x_{opt} versus the PR (β_1) of TCC, temperature ratio (τ) of TC, ε of heat exchanger, ε_R of regenerator, as well as fuel constraint $\overline{Q}_{f*}$, respectively. According to these figures, it is manifest that η_1 can be twice maximized ($\eta_{1max,2}$) at the optimal value (β_{1opt}) of β_1. Besides, η_{1max} exhibits an increasing trend as τ, ε, and ε_R increase, while it exhibits a decreasing trend as $\overline{Q}_{f*}$ increases. One can also see that as β_1 increases, $(\psi_{1opt})_\eta$ first decreases and then increases. However, $(\psi_{1opt})_\eta$ always increases as τ, ε, ε_R, and $\overline{Q}_{f*}$ increase. It is shown that x_{opt} exhibits an increasing trend as β_1 and $\overline{Q}_{f*}$ increase, while exhibits a decreasing trend as τ, ε, and ε_R increase. In addition, one can also note that $(P_{8opt})_\eta$ increases as τ, ε_R, and $\overline{Q}_{f*}$ increase, while it decreases as β_1 and ε increase.

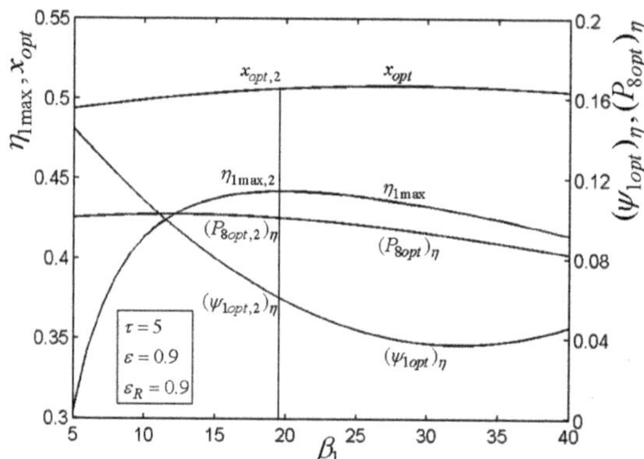

Figure 15. Relationships of $\eta_{1\max} - \beta_1$, $x_{opt} - \beta_1$, $(\psi_{1opt})_\eta - \beta_1$, and $(P_{8opt})_\eta - \beta_1$.

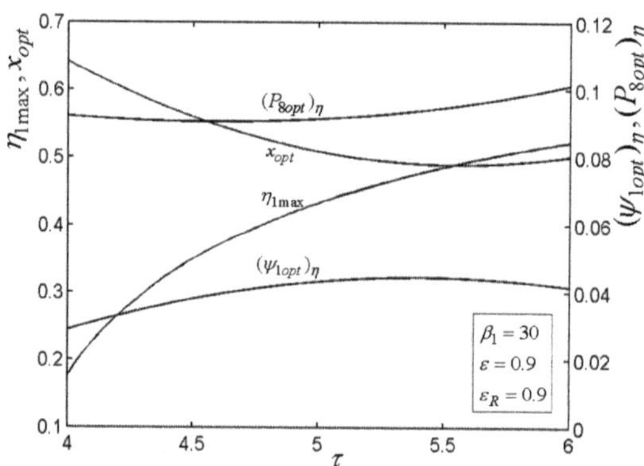

Figure 16. Relationships of $\eta_{1\max} - \tau$, $x_{opt} - \tau$, $(\psi_{1opt})_\eta - \tau$, and $(P_{8opt})_\eta - \tau$.

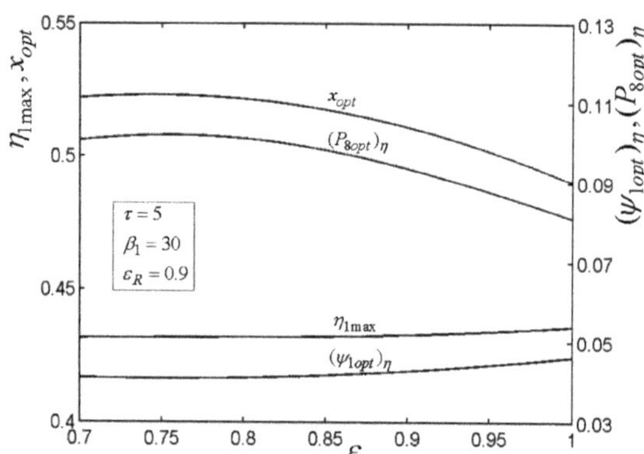

Figure 17. Relationships of $\eta_{1\max} - \varepsilon$, $x_{opt} - \varepsilon$, $(\psi_{1opt})_\eta - \varepsilon$, and $(P_{8opt})_\eta - \varepsilon$.

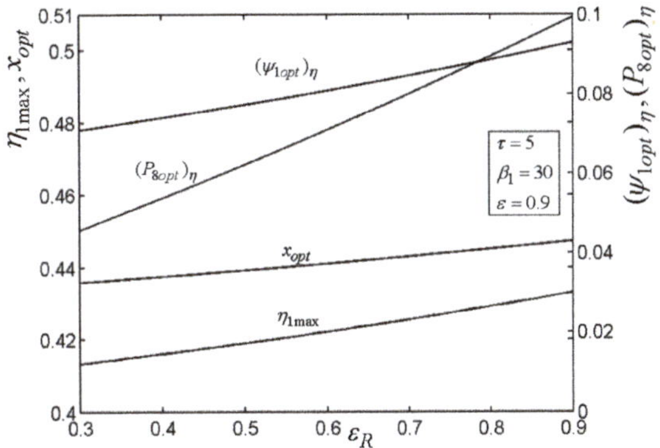

Figure 18. Relationships of $\eta_{1\max} - \varepsilon_R$, $x_{opt} - \varepsilon_R$, $(\psi_{1opt})_\eta - \varepsilon_R$, and $(P_{8opt})_\eta - \varepsilon_R$.

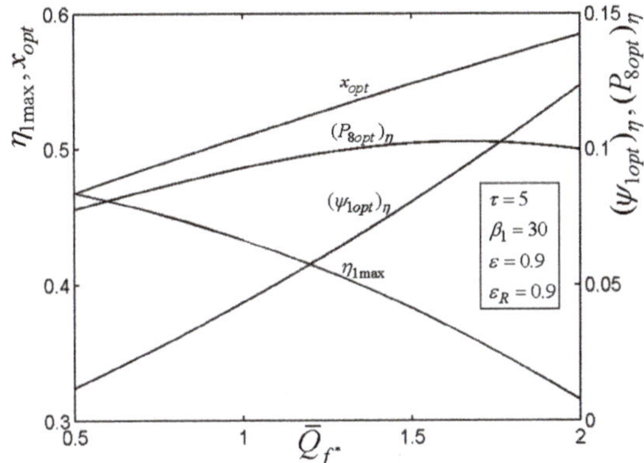

Figure 19. Relationships of $\eta_{1\max} - \overline{Q}_{f*}$, $x_{opt} - \overline{Q}_{f*}$, $(\psi_{1opt})_\eta - \overline{Q}_{f*}$, and $(P_{8opt})_\eta - \overline{Q}_{f*}$.

5. Conclusions

In order to meet the increased request to the effective thermodynamic cycles, more and more new cycle models have been proposed in recently years. Agnew et al. [8] proposed combined Brayton and inverse Brayton cycles in 2003. Based on the combined Brayton and inverse Brayton cycles, Alabdoadaim et al. [9–11] proposed its developed configurations including regenerative cycle and reheat cycle and using two parallel inverse Brayton cycles as bottom cycles. The model cycle discussed in this paper was proposed by Alabdoadaim et al. [11] in 2006. They found that the regenerative combined cycle obtains higher thermal efficiency than that of the base combined cycle but smaller power output at small compressor inlet relative pressure drop of the top cycle based on the first law analysis. Chen et al. [1] established FTT model for this model cycle. This paper is to study the FTT performance in depth. Based on the OCBC model in Ref. [1], performance optimizations of the OCBC are conducted by means of varying the PR of TCC, MFR, as well as PDL allocation in this paper. The maximum PO and TE of the OCBC are gained after optimizations. Furthermore, the influences of cycle parameters on the optimal results are yielded. The numerical results reveal that:

1) Better TE can be procured by introducing the regenerator into the OCBC in contrast with the counterpart without the regenerator put forward by Ref. [7]. However, the performance of PO is inferior in the case of small PD of TCC entrance.

2) The net PO can be maximized by selecting the optimal PD of TCC and PR of BCC. Beyond this, the net PO can be twice maximized at the optimal PR of TCC.

3) The TE can be maximized by selecting the optimal PR of BCC. Additionally, it decreases as the PD of TCC entrance increases.

4) In the premise of constant rate of working fuel and total size of the power plant, TE can be maximized by selecting optimal values of β_i, ψ_1, and x. Furthermore, the TE can be twice maximized by varying the PR of TCC.

5) With consideration of area constraint of the flow cross-sections, TE can be maximized by reasonably selecting the flow areas of the components.

6) There exists optimal PD of TCC entrance. This means that there exist optimal MFR of the working air for the OCBC.

Although the discussed cycle model herein is not validated, the authors of this paper have studied other research objects and partially validated the theoretical models for open Brayton cycles [50,51]. Those can be seen as an illustration for the model herein.

Author Contributions: Conceptualization, L.C.; Funding acquisition, L.C.; Methodology, L.C.; Software, L.C., H.F. and Y.G.; Validation, H.F. and Y.G.; Writing—original draft, L.C.; Writing—review & editing, L.C. All authors have read and agreed to the published version of the manuscript.

Funding: This research was funded by National Natural Science Foundation of China (Grant number 51779262).

Acknowledgments: The authors wish to thank the reviewers for their careful, unbiased, and constructive suggestions, which led to this revised manuscript.

Conflicts of Interest: The authors declare no conflicts of interest.

Nomenclature

A	area
a	cross-section ratio
K	contraction pressure loss coefficient
L	excess air ratio
P	pressure
Q	heat
r	compression ratio
T	temperature
$\overline{W}$	power output
x	area allocation ratio
Greek symbol	
β	pressure ratio
ε	effectiveness
γ	ratio of specific heats
η	efficiency
λ	excessive air ratio
θ	adiabatic temperature ratio
τ	temperature ratio
ψ	pressure drop
Subscripts	
c	compressor
cf	combustor
f	working fuel
g	gas
max	maximum
opt	optimal
R	regenerator
t	turbine
0	ambient
$1, 2, 3, \ldots, 9$	state points in the cycle/sequence numbers

Abbreviations

BCC	bottom cycle compressor
MFR	mass flow rate
OCBC	open combined Brayton cycle
PDL	pressure drop loss
PO	power output
PR	pressure ratio
TCC	top cycle compressor
TE	thermal efficiency

References

1. Chen, L.G.; Zhang, Z.L.; Sun, F.R. Thermodynamic modelling for open combined regenerative Brayton and inverse Brayton cycles with regeneration before the inverse cycle. *Entropy* **2012**, *14*, 58–73. [CrossRef]
2. Radcenco, V.; Vargas, J.V.C.; Bejan, A. Thermodynamics optimization of a gas turbine power plant with pressure drop irreversibilities. *Trans. ASME J. Energy Res. Technol.* **1998**, *120*, 233–240. [CrossRef]
3. Chen, L.G.; Li, Y.; Sun, F.R.; Wu, C. Power optimization of open-cycle regenerator gas-turbine power-plants. *Appl. Energy* **2004**, *78*, 199–218. [CrossRef]
4. Wang, W.H.; Chen, L.G.; Sun, F.R.; Wu, C. Performance optimization of an open-cycle intercooled gas turbine power plant with pressure drop irreversibilities. *J. Energy Inst.* **2008**, *81*, 31–37. [CrossRef]
5. Chen, L.G.; Zhang, W.L.; Sun, F.R. Performance optimization for an open cycle gas turbine power plant with a refrigeration cycle for compressor entrance air cooling. Part 1: Thermodynamic modeling. *Proc. Inst. Mech. Eng. A J. Power Energy* **2009**, *223*, 505–513.
6. Zhang, W.L.; Chen, L.G.; Sun, F.R. Performance optimization for an open cycle gas turbine power plant with a refrigeration cycle for compressor entrance air cooling. Part 2: Power and efficiency optimization. *Proc. Inst. Mech. Eng. A J. Power Energy* **2009**, *223*, 515–522. [CrossRef]
7. Chen, L.G.; Zhang, W.L.; Sun, F.R. Thermodynamic optimization for an open cycle of externally fired micro gas turbine (EFmGT). Part 1: Thermodynamic modeling. *Int. J. Sustain. Energy* **2011**, *30*, 246–256. [CrossRef]
8. Agnew, B.; Anderson, A.; Potts, I.; Frost, T.H.; Alabdoadaim, M.A. Simulation of combined Brayton and inverse Brayton cycles. *Appl. Therm. Eng.* **2003**, *23*, 953–963. [CrossRef]
9. Alabdoadaim, M.A.; Agnew, B.; Alaktiwi, A. Examination of the performance envelope of combined Rankine, Brayton and two parallel inverse Brayton cycles. *Proc. Inst. Mech. Eng. A J. Power Energy* **2004**, *218*, 377–386. [CrossRef]
10. Alabdoadaim, M.A.; Agnew, B.; Potts, I. Examination of the performance of an unconventional combination of Rankine, Brayton and inverse Brayton cycles. *Proc. Inst. Mech. Eng. A J. Power Energy* **2006**, *220*, 305–313. [CrossRef]
11. Alabdoadaim, M.A.; Agnew, B.; Potts, I. Performance analysis of combined Brayton and inverse Brayton cycles and developed configurations. *Appl. Therm. Eng.* **2006**, *26*, 1448–1454. [CrossRef]
12. Zhang, W.L.; Chen, L.G.; Sun, F.R. Power and efficiency optimization for combined Brayton and inverse Brayton cycles. *Appl. Therm. Eng.* **2009**, *29*, 2885–2894. [CrossRef]
13. Zhang, W.L.; Chen, L.G.; Sun, F.R.; Wu, C. Second-law analysis and optimization for combined Brayton and inverse Brayton cycles. *Int. J. Ambient Energy* **2007**, *28*, 15–26. [CrossRef]
14. Chen, L.G.; Zhang, W.L.; Sun, F.R. Power and efficiency optimization for combined Brayton and two parallel inverse Brayton cycles, Part 1: Description and modeling. *Proc. Inst. Mech. Eng. C J. Mech. Eng.* **2008**, *222*, 393–403. [CrossRef]
15. Zhang, W.L.; Chen, L.G.; Sun, F.R. Power and efficiency optimization for combined Brayton and two parallel inverse Brayton cycles, Part 2: Performance optimization. *Proc. Inst. Mech. Eng. C J. Mech. Eng.* **2008**, *222*, 405–414. [CrossRef]
16. Andresen, B. *Finite-Time Thermodynamics*; Physics Laboratory II; University of Copenhagen: Copenhagen, Denmark, 1983.
17. Bejan, A. Entropy generation minimization: The new thermodynamics of finite-size device and finite-time processes. *J. Appl. Phys.* **1996**, *79*, 1191–1218. [CrossRef]

18. Berry, R.S.; Kazakov, V.A.; Sieniutycz, S.; Szwast, Z.; Tsirlin, A.M. *Thermodynamic Optimization of Finite Time Processes*; Wiley: Chichester, UK, 1999.

19. Chen, L.G.; Wu, C.; Sun, F.R. Finite time thermodynamic optimization or entropy generation minimization of energy systems. *J. Non-Equilib. Thermodyn.* **1999**, *24*, 327–359. [CrossRef]

20. Chen, L.G.; Sun, F.R. *Advances in Finite Time Thermodynamics Analysis and Optimization*; Nova Science Publishers: New York, NY, USA, 2004.

21. Feidt, M. Evolution of thermodynamic modelling for three and four heat reservoirs reverse cycle machines: A review and new trends. *Int. J. Refrig.* **2013**, *36*, 8–23. [CrossRef]

22. Ge, Y.L.; Chen, L.G.; Sun, F.R. Progress in finite time thermodynamic studies for internal combustion engine cycles. *Entropy* **2016**, *18*, 139. [CrossRef]

23. Chen, L.G.; Feng, H.J.; Xie, Z.H. Generalized thermodynamic optimization for iron and steel production processes: A theoretical exploration and application cases. *Entropy* **2016**, *18*, 353. [CrossRef]

24. Chen, L.G.; Xia, S.J. *Generalized Thermodynamic Dynamic-Optimization for Irreversible Processes*; Science Press: Beijing, China, 2017.

25. Chen, L.G.; Xia, S.J. *Generalized Thermodynamic Dynamic-Optimization for Irreversible Cycles—Thermodynamic and Chemical Theoretical Cycles*; Science Press: Beijing, China, 2017.

26. Bi, Y.H.; Chen, L.G. *Finite Time Thermodynamic Optimization for Air Heat Pumps*; Science Press: Beijing, China, 2017.

27. Kaushik, S.C.; Tyagi, S.K.; Kumar, P. *Finite Time Thermodynamics of Power and Refrigeration Cycles*; Springer: New York, NY, USA, 2018.

28. Chen, L.G.; Xia, S.J. Progresses in generalized thermodynamic dynamic-optimization of irreversible processes. *Sci. Sin. Technol.* **2019**, *49*, 981–1022. [CrossRef]

29. Chen, L.G.; Xia, S.J.; Feng, H.J. Progress in generalized thermodynamic dynamic-optimization of irreversible cycles. *Sci. Sin. Technol.* **2019**, *49*, 1223–1267. [CrossRef]

30. Chen, L.G.; Li, J. *Thermodynamic Optimization Theory for Two-Heat-Reservoir Cycles*; Science Press: Beijing, China, 2020.

31. Roach, T.N.F.; Salamon, P.; Nulton, J.; Andresen, B.; Felts, B.; Haas, A.; Calhoun, S.; Robinett, N.; Rohwer, F. Application of finite-time and control thermodynamics to biological processes at multiple scales. *J. Non-Equilib. Thermodyn.* **2018**, *43*, 193–210. [CrossRef]

32. Zhu, F.L.; Chen, L.G.; Wang, W.H. Thermodynamic analysis of an irreversible Maisotsenko reciprocating Brayton cycle. *Entropy* **2018**, *20*, 167. [CrossRef]

33. Schwalbe, K.; Hoffmann, K.H. Stochastic Novikov engine with Fourier heat transport. *J. Non-Equilib. Thermodyn.* **2019**, *44*, 417–424. [CrossRef]

34. Fontaine, K.; Yasunaga, T.; Ikegami, Y. OTEC maximum net power output using Carnot cycle and application to simplify heat exchanger selection. *Entropy* **2019**, *21*, 1143. [CrossRef]

35. Feidt, M.; Costea, M. Progress in Carnot and Chambadal modeling of thermomechnical engine by considering entropy and heat transfer entropy. *Entropy* **2019**, *21*, 1232. [CrossRef]

36. Masser, R.; Hoffmann, K.H. Dissipative endoreversible engine with given efficiency. *Entropy* **2019**, *21*, 1117. [CrossRef]

37. Yasunaga, T.; Ikegami, Y. Finite-time thermodynamic model for evaluating heat engines in ocean thermal energy conversion. *Entropy* **2020**, *22*, 211. [CrossRef]

38. Masser, R.; Hoffmann, K.H. Endoreversible modeling of a hydraulic recuperation system. *Entropy* **2020**, *22*, 383. [CrossRef]

39. Chen, L.; Ge, Y.; Liu, C.; Feng, H.J.; Lorenzini, G. Performance of universal reciprocating heat-engine cycle with variable specific heats ratio of working fluid. *Entropy* **2020**, *22*, 397. [CrossRef]

40. Meng, Z.W.; Chen, L.G.; Wu, F. Optimal power and efficiency of multi-stage endoreversible quantum Carnot heat engine with harmonic oscillators at the classical limit. *Entropy* **2020**, *22*, 457. [CrossRef]

41. Bejan, A. *Entropy Generation through Heat and Fluid Flow*; Wiley: New York, NY, USA, 1982.

42. Radcenco, V. *Generalized Thermodynamics*; Editura Techica: Bucharest, Romania, 1994.

43. Bejan, A. Maximum power from fluid flow. *Int. J. Heat Mass Transf.* **1996**, *39*, 1175–1181. [CrossRef]

44. Bejan, A. *Entropy Generation Minimization*; CRC Press: Boca Raton, FL, USA, 1996.

45. Chen, L.G.; Wu, C.; Sun, F.R.; Yu, J. Performance characteristic of fluid flow converters. *J. Energy Inst.* **1998**, *71*, 209–215.

46. Chen, L.G.; Bi, Y.H.; Wu, C. Influence of nonlinear flow resistance relation on the power and efficiency from fluid flow. *J. Phys. D Appl. Phys.* **1999**, *32*, 1346–1349. [CrossRef]

47. Hu, W.Q.; Chen, J.C. General performance characteristics and optimum criteria of an irreversible fluid flow system. *J. Phys. D Appl. Phys.* **2006**, *39*, 993–997. [CrossRef]

48. Radcenco, V. *Optimzation Criteria for Irreversible Thermal Processes*; Editura Tehnica: Bucharest, Romania, 1979.

49. Brown, A.; Jubran, B.A.; Martin, B.W. Coolant optimization of a gas-turbine engine. *Proc. Inst. Mech. Eng. A* **1993**, *207*, 31–47. [CrossRef]

50. Chen, L.G.; Shen, J.F.; Ge, Y.L.; Wu, Z.X.; Wang, W.H.; Zhu, F.L.; Feng, H.J. Power and efficiency optimization of open Maisotsenko-Brayton cycle and performance comparison with traditional open regenerated Brayton cycle. *Energy Convers. Manag.* **2020**, *217*, 113001. [CrossRef]

51. Chen, L.G.; Yang, B.; Feng, H.J.; Ge, Y.L.; Xia, S.J. Performance optimization of an open simple-cycle gas turbine combined cooling, heating andpower plant driven by basic oxygen furnace gas in China's steelmaking plants. *Energy* **2020**, *203*, 117791. [CrossRef]

Article

Chemical and Mechanical Aspect of Entropy-Exergy Relationship

Pierfrancesco Palazzo

Technip Energies, 00148 Roma, Italy; pierfrancesco.palazzo@technipenergies.com or pierfrancesco.palazzo@gmail.com

Abstract: The present research focuses the chemical aspect of entropy and exergy properties. This research represents the complement of a previous treatise already published and constitutes a set of concepts and definitions relating to the entropy–exergy relationship overarching thermal, chemical and mechanical aspects. The extended perspective here proposed aims at embracing physical and chemical disciplines, describing macroscopic or microscopic systems characterized in the domain of industrial engineering and biotechnologies. The definition of chemical exergy, based on the Carnot chemical cycle, is complementary to the definition of thermal exergy expressed by means of the Carnot thermal cycle. These properties further prove that the mechanical exergy is an additional contribution to the generalized exergy to be accounted for in any equilibrium or non-equilibrium phenomena. The objective is to evaluate all interactions between the internal system and external environment, as well as performances in energy transduction processes.

Keywords: Carnot cycle; Carnot efficiency; thermal entropy; chemical entropy; mechanical entropy; thermal exergy; chemical exergy; mechanical exergy; metabolic reactions

Citation: Palazzo, P. Chemical and Mechanical Aspect of Entropy-Exergy Relationship. *Entropy* **2021**, *23*, 972. https://doi.org/10.3390/e23080972

Academic Editor: Michel Feidt

Received: 23 June 2021
Accepted: 19 July 2021
Published: 28 July 2021

Publisher's Note: MDPI stays neutral with regard to jurisdictional claims in published maps and institutional affiliations.

1. Introduction

The research here present follows, and is complementary to, a previous treatise already published and entitled "Thermal and Mechanical Aspect of Entropy-Exergy Relationship" [1]. The purpose is to further extend the perspective already adopted to provide an overarching generalization to include chemical systems and phenomena: in particular, an extension to biological molecules, and molecular aggregates, represents the basis to demonstrate the rigorous and reliable analysis of the relationship between entropy and exergy properties and their applications to chemical non-living and living systems. The interest of such an extension relies in the fact that design and experimental analyses and verifications in different fields of application require implementation of extrema principles based on entropy and exergy as non-conservative and additive state properties. Indeed, in non-equilibrium phenomena, maximum or minimum entropy generation (at macroscopic level) or production (at microscopic level) constitute a methodological paradigm implied in the exergy property founded on the very entropy–exergy relationship. Though, exergy property provides a more complete evaluation of processes since it accounts for: (i) reversible non-dissipative conversions among different forms of energy; and (ii) irreversible dissipative conversions determining entropy creation and related exergy destruction. This dual meaning completeness of the exergy property suggests the research of extrema principles in terms of maximum or minimum exergy disgregation (at macroscopic level) and maximum or minimum exergy degradation (at microscopic level). This extension in turn requests a generalization of properties and processes to chemical internal energy in addition to thermal internal energy usually focused to provide demonstrations and applications of exergy property definitions and exergy method applications. Specific reference is made to the school of thought developed at MIT and reported in publications, textbooks and papers, duly mentioned to describe the paradigm of the methodology as well as the conceptual framework of thermodynamics foundations [2].

139

As it is used in many different contexts and dissertations, it is worth clarifying that the term "generalized" is here used to refer to thermal, chemical and mechanical (and electro-magnetic) aspects of systems and phenomena pertaining to the domain of physical, chemical and biological foundations and applications [3,4].

Moreover, it is worth positing a caveat relating to the concept of heat, mass and work interactions characterizing processes among systems. Indeed, heat, mass and work represent thermal, chemical and mechanical energy transfer and the use of this concept introduces a logical loop in the definition of thermodynamic properties that has been overcome by means of a different set of definitions, assumptions and theorems [5,6]. Despite the use of terms such as "heat or mass or work interaction" should be avoided for this reason, though it is adopted here only to address thermal, chemical and mechanical energy flows and exchanges among systems. In particular, the mass interaction is the homology of heat interaction, whereby particles' potential energy, in terms of chemical potential, is transmitted between two interacting systems, instead of particles' kinetic energy transmitted in the form of heat interaction. In finite terms, the mass interaction occurs through mass entering and exiting the system at constant overall mass, implying that chemical potential is the driving force moving chemical energy associated to chemical entropy fluxes. Nevertheless, mass interaction can be obtained with no bulk-flow through the system and by means of stereochemical variations characterized by isomerization of molecules and polymers.

The interest in developing exergy property and the exergetic method has been high-lighted in different domains, spreading from industry, ecology, biology, as reported in the literature [7]. In exergo-economic applications, exergy has even become the central quantity of a theory of exergetic cost [8].

2. Assumptions and Methods

The dualism consisting of the chemical and mechanical aspects of thermodynamic systems and phenomena represents the chemical–mechanical perspective complementary to the thermal–mechanical one. This conceptual symmetry is further analyzed to provide a definition of the components of entropy and exergy properties relating to mass interactions, typically characterizing chemical processes and chemical internal energy transfer, and work interactions, always occurring along the interaction of any system with a thermal–chemical–mechanical reservoir. Again, in this framework the correlation of chemical potential μ (corresponding to temperature) with respect to chemical internal energy U^C, and the correlation of pressure P with respect to mechanical internal energy U^M, constitute an axiomatic schema. This very schema allows to achieve an extended definition of chemical exergy determined by both chemical potential and pressure, both accounted for in terms of difference with respect to the stable equilibrium state of the external reference system (reservoir) state, is considered. A reservoir is posited to be characterized as behaving at constant chemical potentials and constant pressure (in addition to constant temperature) moving along stable equilibrium states [2].

3. Chemical and Mechanical Components of Entropy Property

As a logical implication of the second law, stated in terms of existence and uniqueness of the stable equilibrium state of a system, the definition of entropy property is proved by using the non-existence of perpetual motion machines of the second kind (PMM2) [2]; the definition of entropy is expressed through the difference between energy and available energy, or exergy, times a parameter characterizing the reservoir [2]. This inferential method is valid for both thermal and chemical components contributing to the entropy balance of any system in any state. Hence, chemical entropy S^C, in addition to thermal entropy S^T, constitutes a property determining the overall internal energy content according to the Euler relation as reported in the literature [2]:

$$U = U^T + U^C + U^M = TS + \sum_{i=1}^{r} \mu_i n_i - PV \tag{1}$$

where μ_i and n_i are the chemical potential and the number of moles, respectively, of the *i*-th chemical species for substances composed by *r* chemical species; this relation, applicable to closed or open systems, hence accounting for "permanent" internal system (non-flow) or "transit" external system (bulk-flow) interactions, can be used to argue for a confrontation between two canonical thermodynamic processes, namely, isothermal and isopotential, as described hereafter. The validity of the phase rule is duly considered since it governs the number of independent intensive quantities determining the thermodynamic state of any system: $F = C - P + 2$, where F is the degree of freedom, C is the number of components, or chemical constituents, and P is the number of phases (solid, liquid, vapor, gas).

In case of isothermal reversible or irreversible processes, the temperature is assumed to remain constant while the system undergoes heat interactions and work interactions simultaneously so that $\delta Q = \delta W \Rightarrow dU = 0$; in the general case of systems undergoing physical operations or chemical reactions, the chemical potential may change along the isothermal process; this is the case of physical operations, such as phase changes (liquid-to-vapor evaporation of vapor-to-liquid condensation), or direct and inverse chemical or stereochemical reactions in which constitutional, conformational or configurational molecular changes occur. In all those different types of isothermal processes, the only result is that heat interaction is transformed into work interaction, or vice versa; hence, in general, the system undergoes chemical potential variations, even though no mass interactions occur and contribute, with interactions intensity and system density, to determine the pressure of the system in addition to the temperature that, instead, remains constant as assumed to characterize the process.

In cases where an isopotential reversible or irreversible process of open systems is analyzed, chemical potentials are assumed to remain constant within the internal system. Notwithstanding both physical operations or chemical reactions may occur, the system undergoes mass interaction and work interaction simultaneously so that $\delta M = \delta W \Rightarrow dU = 0$. Indeed, high chemical potential mass input is compensated for by low chemical potential mass output to ensure no variation of the total mass constituting the system and no variation of chemical potential while a portion of input mass chemical potential is transformed into pressure to allow work interaction. Along an isopotential process, the system may undergo temperature variations (e.g., due to compression or expansion of vapors or gases) even though no heat interactions occur; the temperature contributes towards determining the internal pressure of the system; on this basis, the internal energy variation is formulated by means of the total differential of Euler relation and is expressed, in the specific case of isopotential processes, in the following terms:

$$dU = dU^C + dU^M = d\left(\sum_{i=1}^r \mu_i n_i\right) - d(PV) = \delta M + \delta W = 0 \tag{2}$$

as far as the mechanical term appearing in this relation is concerned, it is null because the isopotential process at constant mass implies that the mechanical internal energy of the whole system remains constant:

$$-d(PV) = -PdV - VdP = 0 \tag{3}$$

where, for the general case of an open system undergoing an isopotential process, the equality $PdV = -VdP$ applies.

Without limiting the generality of this approach, the system considered can be an ideal gas and the thermal form of state equation $PV = \overline{R}T$ applies; though, considering the chemical aspect of the internal system, reference can be made to the chemical form of the state equation that is expressed by means of the chemical potential in lieu of temperature [3]; hence the chemical form of state equation $PV = \overline{R}\mu$ [9] is used to infer that, at constant chemical potential $dU^M = -d(PV) = 0$ as a consequence of the definition of isopotential process, and Equation (2) becomes:

$$dU = dU^C = d\left(\sum_{i=1}^r \mu_i n_i\right) = 0 \tag{4}$$

The chemical energy can also be expressed by means of a formulation relating to the chemical potential and the chemical entropy in the same form of thermal energy, that is:

$$\sum_{i=1}^{r} \mu_i n_i = \sum_{i=1}^{r} \mu_i S_i^C \tag{5}$$

Besides an equivalence factor, the chemical entropy is directly related to the number of molecules, or moles, of any chemical species constituting the internal system; hence, the total differential is:

$$d\left(\sum_{i=1}^{r} \mu_i S_i^C\right) = \sum_{i=1}^{r} \mu_i dS_i^C + \sum_{i=1}^{r} S_i^C d\mu_i \tag{6}$$

However, as the process is isopotential, thus behaving at constant μ_i, then the above equation becomes:

$$dU = dU^C = \sum_{i=1}^{r} \mu_i dS_i^C = 0 \text{ that requires } dS_i^C = 0 \tag{7}$$

The mass interaction occurring along an isopotential process can be realized by means of the addition or subtraction of mass determining the total mass variation of the internal system under consideration; in such a process, both physical operations and chemical reactions are allowed to occur: $dS_i^C = 0$ implies $dn_i = 0$ that is valid if, and only if, the total mass remains constant but, on the other side, the total mass itself has to change due to mass interaction characterizing the assumed isopotential process; hence, the total mass should remain constant and should change at the same time along the same isopotential process, thus representing an apparent contradiction. The resolution of this contradiction relies in the physical meaning of chemical entropy that, instead, is to be considered as a total entropy (of chemical origin), including chemical and mechanical contributions due to mass interaction related to chemical potential, and work interaction related to pressure.

In this regard, as far as the mechanical aspect of the isopotential process is concerned, a further argument relates to the adiabatic reversible process (non-heat and non-mass interactions with external system) that, hence, is accomplished at constant chemical entropy and constant thermal entropy while chemical potential and pressure change along the process; according to the following equations:

$$\begin{aligned}
S^C(\mu, V) - S_0^C &= C_n \ln \frac{\mu}{\mu_0} + \overline{R} \ln \frac{V}{V_0} \\
S^C(\mu, P) - S_0^C &= C_P \ln \frac{\mu}{\mu_0} - \overline{R} \ln \frac{P}{P_0}
\end{aligned} \tag{8}$$

The above expressions are obtained from the homologous ones depending on temperatures of the system; the first term of the second member relates to the chemical potential variation due to chemical reactions occurring in the internal system (with inter-particle potential energy variation and no inter-particle kinetic energy variation), and the second term of the second member relates to the mechanical potential, that is to say, pressure variation due to (internal) work interaction; hence, it can be inferred that the chemical entropy variation, associated to mass interaction, is null by definition of non-mass interaction process; therefore, the way chemical entropy remains constant is because of a compensation effect due to the combination of increasing chemical potential and decreasing pressure, or vice versa, in the internal system.

The specific case of an isopotential reversible or irreversible process is typically representative of a system interaction at constant chemical internal energy. This process requires that both chemical internal energy and chemical entropy remain constant since a mass-to-work or work-to-mass conversion occurs isopotentially by definition, i.e., at constant chemical potential (and constant or variable temperature). This operation is determined by equal quantities of mass input and work output, or work input and mass output. Nevertheless, the mass input or output is associated with a transfer of chemical entropy between internal and external system: hence, a transfer of entropy under "chemical" form requires an entropy transformation into "mechanical" form in order to close the total balance of entropy components to zero, as required by Equation (7), reported here again: $dU = dU^C = \sum_{i=1}^{r} \mu_i dS_i^C = 0$ that implies $dS_i^C = 0$; as mechanical internal energy does not undergo any variation as assumed, this mechanical form of entropy is correlated to work output or input and is determined by pressure and volume variations. In this regard, mechanical entropy is an additional component, and is consistent with, the canonical formulation of entropy calculated for any process, including isopotential, implying that chemical entropy is determined solely by the mass interaction. This conclusion is in compliance with the result provided with the same rationale as for an isothermal reversible process and is described in the process reported in the homologous procedure already mentioned [1].

The analysis described above demonstrates that entropy, in its more general meaning and characterizing internal energy, is constituted by two different and independent components; the first one is the "chemical entropy" that remains constant along an adiabatic reversible process (usually termed as isoentropic), where, instead, only work interaction occurs; the second one is the "mechanical entropy" that remains constant along an isovolumic reversible process where mass interaction (or heat interaction) only occurs. In addition, it can be posited that entropy property S, appearing in the expression of internal energy $U = TS + \sum_{i=1}^{r} \mu_i n_i - PV$, specifically represents the thermal component, or thermal entropy, out of the overall contribution that, nevertheless, remains consistent with, and does not disprove, the above analysis. From a methodological standpoint, the relationship between entropy and exergy properties represents the basis for assuming and proving that chemical and mechanical components set forth for entropy remain valid for chemical exergy and mechanical exergy, respectively.

4. Chemical Exergy Derived from Carnot Chemical Direct Cycle

The definition of chemical exergy analyzed here, among others reported in the literature [10], is based on mass and work interactions and addresses the chemical aspect as a symmetric concept with respect to thermal aspects in the consideration of internal energy contributions. In terms of interactions with the reservoir, the chemical exergy is formulated as the maximum theoretical net useful work withdrawn as a portion of the internal energy of the system, constituting the available energy, along a process leading the system-reservoir composite to the stable equilibrium state. This useful work is calculated on the basis of thermodynamic efficiency of the Carnot chemical direct cycle operating between the variable chemical potential μ of a system A, and the constant chemical potential μ_R of a reservoir R assumed as the external reference system:

$$dEX^C = \delta W_{REV}^{NET} = \delta W_{REV}^{CONVER} + \delta W_{REV}^{TRANSF} \tag{9}$$

where the differential form of chemical exergy is expressed by means of the sum of two terms: (i) a first contribution δW_{REV}^{CONVER} deriving from the conversion of mass interaction into work interaction through a mass-to-work Carnot chemical direct cyclic process [11,12] converting the chemical energy, available at higher chemical potential μ^{HC}, by means of an ideal cyclic machinery operating between μ^{HC} and the reservoir at μ_R^{LC}; (ii) a second contribution δW_{REV}^{TRANSF} deriving from the transfer of mechanical energy by means of work interaction through a cyclic process resulting from system volume variation by means of an ideal machinery operating between P^{HP} and the reservoir at P_R^{LP}; for sake of generality, mass and work interactions can occur either sequentially in different processes or concurrently within the same process; both result from the generalized available energy of a simple system as defined in the approach by Gyftopoulos and Beretta [2].

The rationale to define chemical exergy is based on the confrontation of thermal and chemical aspect of cyclic processes. Usually, temperature is the intensive property determining the Carnot cycle representing the highest efficiency cyclic process and constituting the consequence of the non-existence of perpetual motion machine of the second kind (PMM2) [13]. However, if the same Carnot cycle is regarded as characterized by the chemical potential as an intensive property, instead of temperature, then the Carnot chemical cycle constitutes the symmetric process of a Carnot thermal cycle, considering pressure as the common reference [13]. Hence, based on the chemical potential, a chemical machine model can be described in terms of a chemical conversion cyclic process as the homology of a thermal conversion cyclic process for which balances and efficiencies can be stated [11,12]. In this sense, the above equation, expressing the chemical exergy in differential terms, can be reformulated in the following form:

$$
\begin{aligned}
dEX^C = \delta W_{REV}^{NET} &= \eta_{id}^{CARNOT-CHEMICAL-DIRECT} \cdot \delta M^{HC} + \delta W_{REV}^{TRANSF} \\
&= \frac{\delta W}{\delta M_{ISOPOTENTIAL}^{HC}} \cdot \delta M^{HC} - PdV + P_R dV \\
&= \left(1 - \frac{\mu_R}{\mu}\right) \cdot \delta M^{HC} + \left(1 - \frac{P_R}{P}\right) \cdot \delta W^{HP}
\end{aligned}
\tag{10}
$$

where $\delta M_{ISOPOTENTIAL}^{HC}$ represents the infinitesimal mass interaction along the process at higher chemical potential μ different from the chemical potential μ_R of the reservoir; δM^{HC} represents the infinitesimal mass interaction along any process for which chemical exergy is calculated; δW^{HP} is the infinitesimal work interaction at (variable) high pressure P alongside the process, higher (or lower) with respect to the reservoir (constant) pressure P_R; and the two terms in the last member of the above equation are the consequence of the role of pressure corresponding to the role of chemical potential with respect to mass in chemical exergy.

The above equation is similar to the already known canonical definition of physical exergy [14–16]; this expression is used to define the exergy that is identified by the superscript "C", standing for "Chemical", according to the definition reported in the literature [13] as pointed out above.

In finite terms, considering that $\delta W^{HP} = -PdV$:

$$
\begin{aligned}
EX^C &= W_{10} \\
&= \int_0^1 \left(1 - \frac{\mu_R}{\mu}\right) \cdot \delta M^{HC} + \int_0^1 \left(1 - \frac{P_R}{P}\right) \cdot \delta W^{HP} \\
&= M_{10}^{HC} - \mu_R \int_0^1 \frac{d\mu^{HC}}{\mu^{HC}} + W_{10}^{HP} + P_R \cdot (V_1 - V_0)
\end{aligned}
\tag{11}
$$

where W_{10} is the maximum theoretical net useful work output extracted from the generalized available energy as results from the interaction between system and reservoir; M_{10}^{HC} is the mass interaction alongside the process from the higher isopotential process at μ to the lower isopotential process at μ_R (as a particular case, mass interaction can occur alongside an isopotential process); and W_{10}^{HP} is again the work interaction from the higher isopotential process at μ to the lower isopotential process at μ_R. This equation expresses the chemical exergy EX^C in finite terms as the sum of contributions deriving from cyclic processes where the first one is a mass-to-work ideal cyclic conversion and the second one is an HP-work-to-LP-work ideal cyclic transformation.

The sum of M_{10}^{HC} and W_{10}^{HP} can also be expressed by integrating the Equation (2):

$$
M_{10}^{HC} + W_{10}^{HP} = U_1 - U_0 = M_V \cdot (\mu_1 - \mu_0)
\tag{12}
$$

where the equivalence represents the amount of mass interaction only in the isovolumic processes connecting two states at different chemical potentials. Therefore, chemical exergy can also be associated to sequential isovolumic-isopotential processes connecting any state 1 with a different state 0 of the system. The integral operation results in the expression of chemical exergy, Equation (11). In infinitesimal terms, it constitutes the definition of entropy according to the canonical formulation or, as here proposed, the chemical component of entropy property identified by the superscript "chemical"; the expression in finite terms becomes:

$$
EX^C = W_{10} = (U_1 - U_0) - \mu_R \cdot \left(S_1^C - S_0^C\right) + P_R \cdot (V_1 - V_0)
\tag{13}
$$

where the system-reservoir composite interaction at constant chemical potential μ_R and constant pressure P_R of the reservoir results. This formulation does not contradict the homologous one, proposed by Gyftopoulos and Beretta, deduced from the definition of generalized available energy with respect to an external reference system at constant chemical potential μ_R and constant pressure P_R behaving as a reference external system.

5. Mechanical Exergy Derived from Carnot Chemical Inverse Cycle

The correlation between chemical entropy and chemical exergy clarified in the previous sections is the basis to analyze the entropy–exergy relationship. This analysis is carried out starting from a mechanical standpoint to develop the concept of exergy related, in this case, to work and pressure. To do so, the existence of the mechanical component of entropy already proved is taken into consideration. This different standpoint is viable because the equality of pressure between system and reservoir is an additional condition of the existence and uniqueness of stable equilibrium states of the system-reservoir composite, other than the equality of chemical potential. Indeed, both pressure and chemical potential are thermodynamic potentials driving any equilibrium or non-equilibrium process in the direction of stable equilibrium.

The definition of exergy formulated by the Carnot chemical direct cycle consists of chemical exergy which highlights the role of chemical potential in mass-to-work conversions. On this basis, the research for a definition of mechanical exergy expressed by the inverse cycle is the logical consequence. The objective becomes the physical meaning of the pressure in the opposite process, that is work-to-mass conversion. For the mechanical standpoint too, the general formulation of exergy, in infinitesimal terms, derives from the relationship founded on the Carnot chemical (inverse) cycle and the related expression of thermodynamic efficiency determined by chemical potentials of system and reservoir.

As far as the Carnot cycle is concerned, the usual expression of its performance in terms of thermodynamic efficiency is related to high temperature and low temperature isothermal processes through heat interactions with two reservoirs. Though, the thermodynamic potential constituted by the temperature, or by the inter-particle kinetic energy within the internal system, is continuously transformed into inter-particle potential energy constituting the chemical potential of molecules. In turn, the chemical potential constitutes a thermodynamic potential determining the performance of such a chemical cyclic process. Focusing the performance of chemical cyclic process, it is expressed by means of homologous expression as thermal cyclic processes. Hence, if reference is made to high and low chemical potentials defined as μ^{HC} and μ^{LC} characterizing isopotential processes of the "Carnot chemical cycle", then the formulation of ideal cycle efficiency η_{id}^C is stated as:

$$\eta_{id}^C = 1 - \frac{\mu^{LC}}{\mu^{HC}} \tag{14}$$

These isopotential processes are intended to be characterized by mass interaction input and work interaction output, and vice versa, while the chemical potential of the mass constituting the system remains constant: this means that entering mass implies reducing chemical potential due to chemical reactions occurring at constant temperature while work is exiting the system.

The chemical Carnot cycle considered here will be used to define chemical exergy on the basis of its homology with the canonical thermal Carnot cycle usually referred to in the literature. This chemical cycle, elaborated through ideal processes, is symmetric because it consists of four elaborations, each pair of which is of the same type (isodiabatic), as represented in Figure 1. In case the operating internal system is a perfect gas as assumed, the alternating polytropic processes (two adiabatic and two isopotential), behave according to the following property:

$$\frac{V_1}{V_0} = \frac{V_{1C}}{V_{0C}} \; ; \; \frac{P_1}{P_0} = \frac{P_{1C}}{P_{0C}} \; ; \; \frac{\mu_1}{\mu_0} = \frac{\mu_{1C}}{\mu_{0C}} \tag{15}$$

where the meaning of these ratios is that properties at the end of isodiabatic processes are proportional, therefore the amount of work interaction between internal and external system is the same for both adiabatic compression from 0 to 1 and expansion from 1C to 0C processes; this amount of work interaction is calculated by means of the following expression:

$$W = \frac{1}{K-1} P_0 V_0 \left[\left(\frac{P_1}{P_0} \right)^{\frac{K-1}{K}} - 1 \right] \tag{16}$$

where this depends on the equality $\frac{P_1}{P_0} = \frac{P_{1C}}{P_{0C}}$ and therefore input and output W (with different sign) is equal for the two adiabatic reversible processes. The resulting work interaction balance along the whole cycle accounts for the algebraic sum of work interaction contributions due to both isopotential processes only where mass and work interactions are exchanged simultaneously in directly proportional and equal amounts. This property enables expression of the thermodynamic efficiency of the Carnot chemical cycle of an open bulk-flow system both in terms of mass interaction or work interaction. That thermodynamic efficiency can be expressed either in terms of mass interaction only or in terms of work interaction only due to the equality of mass-work input-output, or vice versa, alongside the isopotential processes as represented in Figure 1:

$$\eta_{id}^{C-DIR} = \frac{W}{M^{HC}} = \frac{W^{HP} - W^{LP}}{M^{HC}} = \frac{W^{HP} - W^{LP}}{W^{HP}} = \frac{M^{HC} - M^{LC}}{M^{HC}} \tag{17}$$

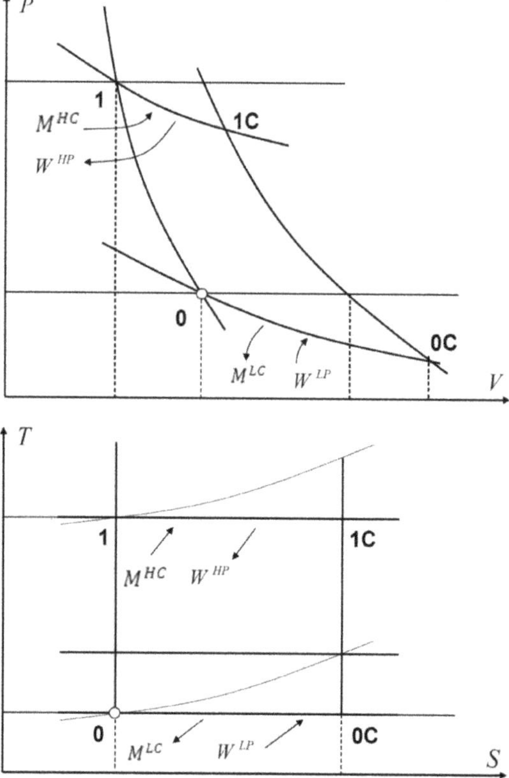

Figure 1. Carnot Chemical Cycle.

As far as the inverse cycle is concerned, if the role of used mass interaction at high chemical potential M^{HC} and utilized total work interaction M^{HC} are replaced by used work interaction W^{HP} and utilized total mass interaction M, the following expression applies:

$$\eta_{id}^{C-INV} = \frac{M}{W^{HP}} = \frac{M^{HC} - M^{LC}}{W^{HP}} = \frac{M^{HC} - M^{LC}}{M^{HC}} = \frac{W^{HP} - W^{LP}}{W^{HP}} = \eta_{id}^{C-DIR} \tag{18}$$

where this equation, for a Carnot chemical inverse cycle, is obtained by assuming the meaning of used and utilized interactions with proper input and output: used mass M^{HC} in the direct cycle corresponds to the used work W^{HP} in the inverse cycle; moreover, the utilized total work W in the direct cycle corresponds to the utilized total mass M in the inverse cycle; as a consequence, the efficiency of a Carnot chemical direct cycle, depending on isopotential processes only, remains unchanged if the Carnot chemical inverse cycle is considered with the corresponding opposite processes; hence the following equality is demonstrated:

$$\eta_{id}^{C-INV} = \frac{M}{W^{HP}} = \frac{W}{W^{HC}} = 1 - \frac{\mu^{LC}}{\mu^{HC}} = \eta_{id}^{C-DIR} \tag{19}$$

It is noteworthy that this approach focuses the concept of exergy and its definitions in terms of used and utilized quantities; thus, it is different from the concept of coefficient of performance (CoP) adopted for refrigeration and cryogenic processes for which used and utilized flows are different and in compliance with operative performances in applications in machinery and plants.

The definition of exergy based on the direct cycle as chemical exergy which is determined by chemical potential in mass-to-work conversion, can be complemented by the symmetric definition of mechanical exergy founded on the inverse cycle; in this case, the physical meaning of pressure in the opposite work-to-mass conversion, determines the pressure level of work interactions alongside the higher chemical potential, and higher pressure, isopotential processes of the Carnot chemical inverse cycle.

The concept of equivalence and interconvertibility, demonstrated by Gaggioli [14–16], can be stated in different terms: "useful work is not better than useful mass, and the available energy results in maximum net useful mass or, equivalently, maximum net useful work, or the combination of both." Thus, the definition of mechanical exergy representing, in this case, the maximum net useful mass withdrawable from the available energy, in infinitesimal terms, can be expressed as:

$$dEX^M = \delta M_{REV}^{NET} = \delta M_{REV}^{CONVER} + \delta M_{REV}^{TRANSF} \tag{20}$$

where the first term of the last member δM_{REV}^{CONVER} is the net amount of mass interaction resulting from the balance of a Carnot chemical inverse cycle converting the available work at pressure P into mass through the interaction with a reservoir at constant pressure P_R; the second term of the last member δM_{REV}^{TRANSF} is the net amount of available energy transferred from the external to the internal system by means of mass interaction alongside a non-cyclic or cyclic process; mass and work interactions are accounted for occurring either successively or simultaneously, and both derive from the generalized available energy of a simple system as defined by Gyftopoulos and Beretta [2]. Hence, in differential terms:

$$dEX^M = \delta M_{REV}^{NET} = \eta_{id}^{CHEMICAL-CARNOT-INVERSE} \cdot \delta W^{HP} + \delta M_{REV}^{TRANSF} \tag{21}$$

On the basis of Equation (15):

$$\begin{aligned} dEX^M &= \frac{\delta M}{\delta W_{ISOTHERMAL}^{HP}} \cdot \delta W^{HP} + \mu dS^C - \mu_R dS^C \\ &= \frac{\delta M}{\delta W_{ISOTHERMAL}^{HP}} \cdot \delta W^{HP} + (\mu - \mu_R)dS^C \\ &= \left(1 - \frac{\mu_R}{\mu}\right) \cdot \delta W^{HP} + \left(1 - \frac{\mu_R}{\mu}\right) dM^{HC} \end{aligned} \tag{22}$$

The formulation of chemical exergy is now reversed to define the mechanical exergy, identified by the superscript "M", that is not related to exergy associated to center-of-mass macroscopic kinetic and potential energy, already termed as "kinetic exergy" and "potential exergy" according to the literature.

After replacing work with mass, the mechanical exergy in finite terms is formulated as:

$$EX^M = M_{10} = \int_0^1 \left(1 - \frac{\mu_R}{\mu}\right) \cdot \delta W^{HP} + \int_0^1 (\mu - \mu_R) \cdot dS^C$$
$$= W_{10}^{HP} - \mu_R \int_0^1 \frac{\delta W^{HP}}{\mu} + M_{10}^{HC} - \mu_R \left(S_1^C - S_0^C\right) \tag{23}$$

where M_{10} is the maximum theoretical net useful mass output obtained by means of the generalized available energy resulting from the interaction process between system and reservoir; W_{10}^{HP} is the work interaction from higher isopotential curve at μ and corresponds with the state at pressure P to lower isopotential curve at μ_R; as a particular case, the work interaction can occur alongside an adiabatic reversible process; the sum of terms W_{10}^{HP} and M_{10}^{HC} in the last member of previous equation can also be expressed as:

$$W_{10}^{HP} + M_{10}^{HC} = U_1 - U_0 = C_V(\mu_1 - \mu_0) \tag{24}$$

This equation expresses the equivalence with the sole amount of work interaction in a chemical (and thermal) isoentropic process (where work interaction only occurs), between two different chemical potentials. Hence, the mechanical exergy characterizes an isoentropic-isopotential sequential process connecting the generic state 1 with the stable equilibrium state 0 of the system–reservoir composite. If the chemical state equation $PV = \overline{R}\mu$ is adopted and used in the expression of mechanical exergy, then:

$$EX^M = Q_{10} = (U_1 - U_0) - \overline{R}\mu_R \int_0^1 \frac{\delta W^{HP}}{PV} - \mu_R \left(S_1^C - S_0^C\right) \tag{25}$$

The integrand term $\frac{\delta W^{HP}}{PV}$ of the above equation is formally homologous of the integrand term $\frac{dM^{HC}}{\mu}$ representing the very definition of chemical entropy according to the concept and the definition of entropy property as per Clausius formulation; on the basis of this formal homology extended to work interaction and the mechanical internal energy, the definition of mechanical entropy is derived and formulated as:

$$dS^M = \frac{\delta W^{HP}}{PV} \tag{26}$$

where the factor $1/PV$ is the integrating factor of the infinitesimal work interaction δW^{HP} that changes the integrand function into an exact differential function; indeed, assuming the expression of mechanical exergy previously reported as Equation (22), and considering that $\delta W^{HP} = -PdV$ then it is allowed to differently express the mechanical exergy (of chemical origin) as:

$$EX^M = M_{10}$$
$$= (U_1 - U_0) - \overline{R}\mu_R \int_0^1 \frac{\delta W^{HP}}{PV} - \mu_R\left(S_1^C - S_0^C\right)$$
$$= (U_1 - U_0) + \mu_R \int_0^1 \overline{R}\frac{dV}{V} - \mu_R\left(S_1^C - S_0^C\right) \tag{27}$$
$$= (U_1 - U_0) + \mu_R\left(\overline{R}\ln V_1 - \overline{R}\ln V_0\right) - \mu_R\left(S_1^C - S_0^C\right)$$

where it relates to the work interaction with the environmental system represented by the mechanical reservoir; therefore, considering that the chemical state equation $PV = \overline{R}\mu$ applies, then:

$$EX^M = M_{10} = (U_1 - U_0) + P_R V_R(\ln V_1 - \ln V_0) - \mu_R\left(S_1^C - S_0^C\right) \tag{28}$$

where the homology with the expression of chemical exergy (and thermal exergy) demonstrates the common origin of all exergy components deriving from conversion processes from one energy form to a different one characterized by entropy variations occurring along those processes; to complete this homology, the integrating factors included in the integration function are similar:

$$dS^T = \frac{\delta Q^{HT}}{T} \text{ similar to } dS^C = \frac{\delta M^{HC}}{\mu} \text{ similar to } dS^M = \frac{\overline{R} dV}{V} \tag{29}$$

where the last differential is integrated, the equation in finite terms becomes:

$$S^M = \overline{R} \ln V + C \tag{30}$$

Hence, dS^M being an exact differential function, then S^M is a state property depending on the state parameter volume and can be adopted as the formal definition of mechanical entropy; moreover, as volume is additive, the mechanical entropy is an additive property. As far as the dimensional analysis is concerned, since the logarithmic function is dimensionless, then the dimension of mechanical entropy is related to $\overline{R}$ having dimensions $\left(J \cdot kg^{-1} \cdot K^{-1} \right)$ or $\left(J \cdot mol^{-1} \cdot K^{-1} \right)$ that are identical to the chemical entropy and thermal entropy dimensions. In this regard, the relationship between mechanical exergy and volume, and pressure as a consequence, constitutes the rationale for considering the equality of pressure between system and reservoir, as an additional condition of mutual stable equilibrium to be accounted for in the definitions of available energy and exergy, and hence in the definition of entropy property related to, and derived from, energy and available energy or exergy according to the proof method demonstrated and reported in the literature. The physical meaning of mechanical exergy can be ascribed to the combination of pressure characterizing the mechanical internal energy of the system, and the pressure of work interaction occurring between system and reservoir. It is noteworthy that the demonstration procedure described here is, in its rationale, identical to the one stated to achieve the mechanical entropy definition based on thermal entropy using the corresponding quantities to replicate the proof.

6. Generalized Chemical Exergy Related to Chemical-Mechanical Reservoir

The definition of chemical entropy and mechanical entropy, derived and expressed from chemical exergy and mechanical exergy, respectively, is accounted for here to generalize the conceptual definition of chemical exergy including mass interaction, in addition to work interaction, characterizing interaction processes occurring between system and reservoir. On the basis of equivalence and interconvertibility proposed by Gaggioli et al. [9,10] for thermal and mechanical aspect of interactions, and here mutuated for chemical and mechanical interactions, the exergy of a system interacting with a reservoir results in the following statements:

(1) Exergy is the available work or maximum theoretical net useful work constituting the chemical exergy;
(2) Exergy is the available mass or maximum theoretical net useful mass constituting the mechanical exergy;

The generalization of chemical exergy proposed here is, for the above rationale, implicated with the chemical exergy underpinned by the Carnot chemical direct cycle efficiency and the high chemical potential mass interaction; chemical exergy additionally contributes to the mechanical exergy underpinned by the Carnot chemical inverse cycle efficiency and the high pressure work interaction; both exergies are defined considering a chemical–mechanical reservoir at constant chemical potential and constant pressure behaving at permanent stable equilibrium according to the canonical definition of reservoir. The generalized chemical exergy outlined above takes into account the implication of pressure in work interaction that generates different amounts of mass interaction depending on different pressure values at which the work interaction occurs. In different terms, it can be stated that the same amount of available mechanical internal energy transferred by means of work interaction can be used at different pressure of the system with respect to the constant pressure of the (mechanical) reservoir, to be converted into mass interaction at different chemical potentials. Hence, the useful work, available in the form of mechanical available energy, is also evaluated in terms of the second law by means of the Carnot chemical inverse cycle, producing the mass interaction output: therefore, mass-to-work conversion and work-to-mass conversions are accounted for simultaneously—this implies that the generalized chemical exergy can be regarded in the perspective of an "exergy of exergy" that makes work interaction equivalent to, and interconvertible with, mass interaction, and vice versa.

Before achieving the formulation of the generalized chemical exergy, the differential form of internal energy in differential terms according to Gibbs' equation is considered:

$$dU = \sum_{i=1}^{r} \mu_i dn_i - PdV = \delta M + \delta W \tag{31}$$

This can be reformulated in different terms by adopting the chemical entropy and the mechanical entropy previously defined and specified for all chemical substances constituting the internal system; this reformulation is a crucial step in the direction of a generalized Gibbs equation that, in this perspective, is modified into the following:

$$dU = \sum_{i=1}^{r} \mu_i dS_i^C - \sum_{i=1}^{r} \frac{P_i V_i}{R} dS_i^M = \delta M + \delta W \tag{32}$$

where, in turn, it can be expressed by means of the chemical state Equation [5]:

$$\begin{aligned} dU &= \sum_{i=1}^{r} \mu_i dS_i^C - \sum_{i=1}^{r} \mu_i dS_i^M \\ &= \sum_{i=1}^{r} \mu_i \left(dS_i^C - dS_i^M \right) = \delta M + \delta W \end{aligned} \tag{33}$$

The term $\left(dS_i^C - dS_i^M \right)$ represents the differential generalized entropy dS_i^G which, in finite terms, is $S_i^G = S_i^C - S_i^M$ associated to, and depending on, the chemical potentials and is determined by mass interaction and work interaction contributing to the variation of the internal energy. Equation (33) above can be expressed as:

$$dU = \sum_{i=1}^{r} \mu_i dS_i^G = \delta M + \delta W \tag{34}$$

where, in finite terms, U being a state property determined by two independent variables, the following generalized Gibbs equation is deduced:

$$U = U(S, V) = \sum_{i=1}^{r} \mu_i \Delta S^G = M + W \tag{35}$$

The generalized entropy is the result of the contribution of chemical and mechanical components and represents the rationale for resolving the apparent inconsistency expressed by the statement: $dU = dU^C = \sum_{i=1}^{r} \mu_i dS_i^C = 0$ implying that $dS_i^C = 0$; indeed, the Gibbs equation is allowed to be null because of the two terms of $S_i^G = S_i^C - S_i^M$, which, in the special case of an isopotential process of a perfect and single-phase homogeneous gas describing the internal system, are expressed as:

$$\Delta S_{ISOPOTENTIAL}^C = C_n \ln \frac{\mu}{\mu_0} + \overline{R} \ln \frac{V}{V_0} \tag{36}$$

$$\Delta S_{ISOPOTENTIAL}^M = \overline{R} \ln \frac{V}{V_0} \tag{37}$$

These two terms used to replace the corresponding ones in the $S_i^G = S_i^C - S_i^M$ become:

$$\Delta S_{ISOPOTENTIAL}^G = \Delta S_{ISOPOTENTIAL}^C - \Delta S_{ISOPOTENTIAL}^M = C_n \ln \frac{\mu}{\mu_0} = 0 \tag{38}$$

confirming that $\Delta U = 0$ for an isopotential reversible or irreversible process as $\Delta S^G_{ISOPOTENTIAL} = 0$ as required to resolve the inconsistency of conditions $dU = dU^C = \sum_{i=1}^{r} \mu_i dS^C_i = 0$ implying $dS^C_i = 0$ before positing.

As far as isovolumic processes are concerned, the same approach is applied by evaluating the two components of generalized entropy along the process:

$$\Delta S^C_{ISOVOLUMIC} = C_n \ln \frac{\mu}{\mu_0} \tag{39}$$

$$\Delta S^M_{ISOVOLUMIC} = 0 \tag{40}$$

Thus, the sum of the two contributions is:

$$\Delta S^G_{ISOVOLUMIC} = \Delta S^C_{ISOVOLUMIC} - \Delta S^M_{ISOVOLUMIC} = C_n \ln \frac{\mu}{\mu_0} \tag{41}$$

Therefore, the generalized entropy is identical to the chemical entropy, that confirming the dependence on the chemical potential as the overall and unique thermodynamic potential determining the state of the system.

In case of an isobaric process, the following applies:

$$\Delta S^C_{ISOBARIC} = C_P \ln \frac{\mu}{\mu_0} - \overline{R} \ln \frac{P}{P_0} \tag{42}$$

$$\Delta S^M_{ISOBARIC} = \overline{R} \ln \frac{V}{V_0} \tag{43}$$

Again, the sum of the two contributions is:

$$\begin{aligned}
\Delta S^G_{ISOBARIC} &= \Delta S^C_{ISOBARIC} - \Delta S^M_{ISOBARIC} \\
&= C_P \ln \frac{\mu}{\mu_0} - \overline{R} \ln \frac{P}{P_0} - \overline{R} \ln \frac{V}{V_0} \\
&= C_n \ln \frac{\mu}{\mu_0} + \overline{R} \ln \frac{V}{V_0} - \overline{R} \ln \frac{V}{V_0} \\
&= C_n \ln \frac{\mu}{\mu_0}
\end{aligned} \tag{44}$$

Finally, for an adiabatic reversible process:

$$\Delta S^C_{ADIABATIC} = 0 \tag{45}$$

$$\Delta S^M_{ADIABATIC} = \overline{R} \ln \frac{V}{V_0} \tag{46}$$

$$\begin{aligned}
\Delta S^G_{ADIABATIC} &= \Delta S^C_{ADIABATIC} - \Delta S^M_{ADIABATIC} = 0 - \Delta S^M_{ADIABATIC} \\
&= -\overline{R} \ln \frac{V}{V_0} = C_n \ln \frac{\mu}{\mu_0}
\end{aligned} \tag{47}$$

hence demonstrating, by means of Equations (8), the existence of the relationship between pressure, that changes with volume, and the generalized entropy in the special case of absence of mass interaction determining chemical entropy null variations.

To summarize, a first outcome is that the method applied to explain the mechanical entropy contribution has led to resolve the apparent controversy already mentioned and provides a formal definition of mechanical entropy related to the pressure, with a direct implication with the definition of mechanical exergy property. A second outcome, deriving from the above method, concerns the dependence of the generalized chemical entropy solely on the chemical potential in all thermodynamic processes analyzed above; this outcome can be derived from the physical meaning of internal energy pertaining to a real, multi-phase, non-homogeneous, internal system characterized by atomic-molecular chemical bonds and interactions regardless of the thermal state and heat interactions between internal and external systems.

A caveat concerning the assumption that the canonical processes above are not limited to reversible conditions, irreversible processes are accounted for.

That said, on the basis of the relationship between generalized entropy and internal energy, if the external system behaves as a chemical and mechanical reservoir in stable equilibrium state at constant chemical potential and pressure, the internal energy balance of the system-reservoir composite is expressed as:

$$
\begin{aligned}
EX^C &= -\left(W^{AR\leftarrow}\right) - \left(M^{AR\leftarrow}\right) \\
&= \Delta U^{SYSTEM} + \Delta U^{RESERVOIR} \\
&= \Delta U_W^{SYSTEM} + \Delta U^{R,W} + \Delta U_M^{SYSTEM} + \Delta U^{R,M}
\end{aligned}
\tag{48}
$$

The conceptual meaning of this expression is that $\Delta U_W^{SYSTEM} + \Delta U^{R,W}$ equals the mechanical exergy converted into chemical exergy, and $\Delta U_M^{SYSTEM} + \Delta U^{R,M}$ equals the chemical exergy converted into mechanical exergy; in different terms:

$$
EX^C = -\left(W^{AR\leftarrow}\right) - \left(M^{AR\leftarrow}\right) = (U - U_0) - M_R - W_R
\tag{49}
$$

where M_R is the minimum mass interaction representing the (minimum) mechanical exergy (Equation (28)) lost to the chemical reservoir and W_R is the minimum work interaction representing the (minimum) chemical exergy (Equation (13)) lost to the mechanical reservoir. The symbol EX^C (or, according to some authors, X^C), in lieu of M and W, is adopted here to identify the chemical exergy generalized in its physical and chemical meaning as deriving from the combination of useful work and useful mass. The arrow in the superscript means that the interaction enters the system, according to the symbology adopted by Gyftopoulos and Beretta [2].

The Carnot cycle represented in Figure 1 constitutes the rationale for the generalized formulation of chemical exergy; indeed, the isopotential process verifies the equality $M^{AR} = W^{AR}$ alongside both high and low chemical potential processes where, instead, the chemical potential is constant but the pressure is not; therefore, W^{AR} at decreasing pressure constitutes an amount of (chemical) exergy that should be considered a loss of mechanical internal energy since it is released isopotentially to the reservoir while chemical internal energy is transferred from the reservoir to the system at stable equilibrium conditions; those isopotential processes are the result of chemical-to-mechanical and mechanical-to-chemical internal energy transformations implying entropy transformation appearing in the equation of generalized chemical exergy:

$$
\begin{aligned}
EX^G &= -\left(W^{AR\leftarrow}\right) - \left(M^{AR\leftarrow}\right) = \Delta U^{SYSTEM} + \Delta U^{RESERVOIR} \\
&= (U_1 - U_0) \quad \text{variation of internal energy of the system} \\
&\quad - \sum_{i=1}^{r} \mu_i \Delta S_i^C \quad \text{energy conversion within the system} \\
&\quad - \sum_{i=1}^{r} \mu_i \Delta S_i^{C,R} \quad \text{chemical energy transfer between system and reservoir} \\
&\quad + P_R \Delta V^R \quad \text{mechanical energy transfer system} - \text{to} - \text{reservoir}
\end{aligned}
\tag{50}
$$

It is of crucial importance highlighting that the concept of entropy conversion is inherent to the concept of energy conversion occurring in any cyclic process, and, for this very reason, intrinsic to the concept of exergy; hence, entropy conversion occurring along a cyclic process implies the additional term expressing the contribution of the mechanical component to the overall cycle entropy balance and the subsequent exergy balance representing the basis of a property's efficiency and, finally, the performance quantification. Replacing the expressions of chemical exergy EX^C and mechanical exergy EX^M in the above equation of generalized exergy EX^G, the following equation is derived:

$$
\begin{aligned}
EX^G = \;&(U_1 - U_0) \\
&- \textstyle\sum_{i=1}^{r} \mu_{i,R}\Delta S_i^C \quad &\text{chemical energy conversion loss released to reservoir} \\
&+ P_R \Delta V_R \quad &\text{mechanical energy transformation loss released to reservoir} \\
&+ P_R V_R (\ln V_1 - \ln V_0) \quad &\text{mechanical energy conversion loss released to reservoir} \\
&- \textstyle\sum_{i=1}^{r} \mu_{i,R}\Delta S_i^C \quad &\text{chemical energy transformation loss released to reservoir}
\end{aligned}
\tag{51}
$$

The term $P_R V_R (\ln V_1 - \ln V_0)$ constitutes the "entropic-mechanical" component taking into account the entropy conversion undergone by the system along the conversion cycle (and in particular due to adiabatic processes) and representing a contribution, in addition to the chemical entropy, to the overall cycle balance.

As all properties are additive, the generalized chemical exergy can be stated in the following explicit form:

$$
\begin{aligned}
EX^G = \;&-\left(W^{AR\leftarrow}\right) - \left(M^{AR\leftarrow}\right) \\
&+ \left[(U - U_0) - \textstyle\sum_{i=1}^{r} \mu_{i,R}\Delta S_i^C + P_R\Delta V_R\right]^C \\
&+ \left[(U - U_0) + P_R V_R(\ln V_1 - \ln V_0) - \textstyle\sum_{i=1}^{r} \mu_{i,R}\Delta S_i^C\right]^M
\end{aligned}
\tag{52}
$$

where: the first term, that is the first square parenthesis ("chemical") of the second member of Equation (52), is the contribution relating to the variation of internal energy due to the mass interaction corresponding to the chemical exergy; the second term, that is second square parenthesis ("mechanical") of the second member of Equation (52), is the contribution relating to the variation of internal energy due to the work interaction corresponding to the mechanical exergy; both chemical exergy and mechanical exergy constitute the two components of the generalized chemical exergy along any process. Indeed, as the internal energy is an additive state property, both contributions determined by mass interaction or work interaction with the external system (useful of reservoir), can occur sequentially or simultaneously to connect any pair of thermodynamic states. Hence, the first term constitutes the chemical exergy calculated alongside an isovolumic-isopotential process and the second term constitutes the mechanical exergy calculated alongside an isoentropic-isopotential process.

The meaning of the generalized chemical exergy is highlighted for an adiabatic and esoergonic reversible process for which work interaction only characterizes the thermodynamic state and no mass interaction and no heat interaction occur. This process is determined by absence of chemical entropy and thermal entropy variations (due to absence of mass interaction and heat interaction, respectively) and a non-null variation of mechanical entropy (due to occurring work interaction). As a consequence of the generalized formulation, if this adiabatic process is calculated in terms of exergy, the available energy (in the form of pressure mechanical energy withdrawable from the system) is accounted for in terms of its capability to be converted (and not directly transferred) into useful mass; the consequence is that the exergy analysis implies a lower amount if compared with the canonical method that identifies exergy exclusively with work interaction output conveyed to, and used by, the external system, as it is. In this regard, the entropic-mechanical addendum of the generalized chemical exergy, Equations (28) and (52), determines a reduction due to the work interaction undergoing the (reversible) entropy conversion that makes this work input not useful for a work-to-mass conversion into mass output.

7. Outcomes and Applications

The domain of applications of the generalized chemical exergy spreads to inorganic and organic chemistry including metabolic biological processes in living organisms. Metabolic processes determine morphological development and homeostasis as well as energetic transduction in living cells and are subdivided into two main categories: (i) catabolic processes implying the demolition of molecule aggregates and (ii) anabolic processes aimed at building-up proteins, enzymes and other organic substances and precursors. In catabolic processes, such as glycolysis, the chemical energy of glucose is transformed into chemical energy in the form of free enthalpy of adenosine tri-phosphate (ATP) [17]. The glycolysis is subdivided in two phases: (i) storing phase and (ii) releasing phase. The ATP releases chemical energy to the D-glyceraldehide-3-phosphate and is stored in these molecules during the first phase in 5 steps; instead, during the 5 steps of the second phase of glycolysis, the same chemical energy is released back to ATP, NADH and pyruvate which are products of the whole glycolysis catabolic process. The complete glycolysis process encompasses chemical exergy storage and subsequent release and the corresponding mechanical exergy and chemical exergy characterize the bi-directional inverse and direct conversions. Three, out of ten, of these processes (1st, 3rd and 10th), are irreversible and govern the entire series of reactions. The pyruvate undergoes a subsequent aerobiotic oxidation process, followed by the Krebs cycle and ending with the oxidative phosphorylation characterized by the following final oxidation reaction: $NADH + H^+ + 1/2\,O_2 \rightarrow NAD^+ + H_2O$. The ATP is used in multiple anabolic processes and the NADH and $FADH_2$, reduced electron transporters, are involved in several metabolic processes [17]. In particular, the ATP is used by living organisms to release mechanical work interaction as chemical exergy output used for locomotion, for food, recovery, defense, reproduction and all other activities needed for life. Using the generalized chemical exergy provides a method to analyze aggregates, such as amino acids, proteins, enzymes and nucleic acids, constituting molecular machines, non-cyclic or cyclic, characterized by phenomena, balances and efficiencies governed by the microscopic thermodynamics at atomic and molecular level [18–21]. In this perspective, a contribution could arise in the direction of researches focusing metabolic paths and cell membrane role [22]. This approach is in use in various diseases already undergoing studies and experimental investigations [23–25].

8. Conclusions

The main outcome of the procedure described is that the generalized chemical exergy can be expressed by the sum of the two components defined as chemical exergy and mechanical exergy:

$$EX^{CHEMICAL} = EX^C + EX^M = W_{REV}^{CONVER} + W_{REV}^{TRANSF} + M_{REV}^{CONVER} + M_{REV}^{TRANSF} \quad (53)$$

This does not depend on a particular process adopted for its definition; thus, it can be considered as a general formulation which valid for any process, reversible or irreversible, connecting two different thermodynamic states.

Another outcome is that the generalized chemical exergy is determined by the equality of pressure, in addition to the equality of chemical potential, as a further condition of mutual stable equilibrium between system and reservoir. In the perspective of implications of this additional condition and the generalization to any system (large and small) in any state (equilibrium and non-equilibrium), the concept of generalized chemical exergy would require the reference to a mechanical reservoir behaving at constant pressure in addition to the chemical reservoir.

In the framework of the Gyftopoulos and Beretta perspective, the formulation of chemical entropy can be expressed in the following form adopting the symbol E to denote energy and Ω to denote available energy [2]:

$$(S_1 - S_0)^C = \frac{1}{\mu_R}\left[(E_1 - E_0) - \left(\Omega_1^R - \Omega_0^R\right)\right]^C \quad (54)$$

where, if the concept of mechanical reservoir is introduced, and the equality of pressure between the system and the mechanical reservoir is considered as an additional condition for mutual stable equilibrium between system and reservoir, it remains valid. The additive property of entropy would lead to assuming that [2,13]:

$$(S_1 - S_0)^M = \frac{\overline{R}}{P_R V_R}[(E_1 - E_0) - \left(\Omega_1^R - \Omega_0^R\right)]^M \tag{55}$$

where the mechanical component of entropy would be defined with reference to a mechanical reservoir at constant pressure.

Finally, the additivity of entropy components allows stating the following:

$$(S_1 - S_0)^G = (S_1 - S_0)^C + (S_1 - S_0)^M \tag{56}$$

This should be proved to complete the formulation of generalized entropy which takes into account the general definitions proposed for chemical entropy and mechanical entropy.

As a conclusion of the present research, the methodology adopted has achieved a result for the chemical aspect that can be considered homologous to the result in the procedure already adopted for the thermal aspect and mentioned at the outset of this treatise. The equality of chemical potentials, as a condition of mutual stable equilibrium in addition to the equality of temperature and pressure of the composite system-reservoir, is an important result. Hence, the set of all conditions of mutual stable equilibrium enables establishing a more complete formulation of generalized exergy with the contribution of chemical, thermal and mechanical exergy related to a 'thermo-chemical-mechanical' reservoir. As a consequence, the definition of chemical entropy has been derived in relation with the molecular geometry of any system in any state, including non-equilibrium. In consideration of the importance of thermodynamic methods in chemistry and biology [26–29], different studies and applications have been developed focusing extrema principles and constructal laws [30–32]. In this regard, it would be worth thinking and fostering a line of research aimed at building up a rational and systematic paradigm including thermodynamic and informational aspects both constituting of intrinsic fundamentals of systems and phenomena associated with life.

Funding: This research received no external funding.

Conflicts of Interest: The author declares no conflict of interest.

Nomenclature

C	specific heat or specific mass
E	energy
EX	exergy
M	mass interaction
P	pressure
Q	heat interaction
$\overline{R}$	universal constant of gases
S	entropy
T	temperature
U	internal energy
V	volume
W	work interaction

Greek Symbols

μ	chemical potential
Ω	available energy

Superscripts

$\leftarrow$	heat, mass or work interaction entering the system
$\rightarrow$	heat, mass or work interaction exiting the system
AR	composite of interacting system A and reservoir R
C	chemical

CONVER	conversion
DIR	direct
G	generalized
HC	high chemical potential
HP	high pressure
INV	inverse
LC	low chemical potential
LP	low pressure
M	mechanical
T	thermal
TRANSF	transfer
Subscripts	
0	initial state
1	final state
i	*i*-th chemical constituent
n	number of moles
r	number of chemical constituents
R	reservoir
V	constant volume

References

1. Palazzo, P. Thermal and mechanical aspect of entropy-exergy relationship. *Int. J. Energy Environ. Eng.* **2012**, *3*, 4. [CrossRef]
2. Gyftopoulos, E.P.; Beretta, G.P. *Thermodynamics: Foundations and Applications*; Dover Publications: New York, NY, USA, 2005.
3. Demirel, Y.; Sandler, S.I. Nonequilibrium Thermodynamics in Engineering and Science. *J. Phys. Chem. B* **2004**, *108*, 31–43. [CrossRef]
4. Demirel, Y.; Gerbaud, V. *Nonequilibrium Thermodynamics: Transport and Rate Processes in Physical, Chemical and Biological Systems*, 4th ed.; Elsevier: Amsterdam, The Netherland, 2019.
5. Zanchini, E.; Beretta, G.P. Removing Heat and Conceptual Loops from the Definition of Entropy. *Int. J. Thermodyn.* **2010**, *13*, 67–76.
6. Badur, J.; Feidt, M.; Ziolkowsky, P. Without heat and work—Further remarks on the Gyftopoulos-Beretta Exposition of Thermodynamics. *Int. J. Thermodyn.* **2018**, *21*, 180–184. [CrossRef]
7. Sciubba, E.; Wall, G. A Brief Commented History of Exergy from the Beginning to 2004. *Int. J. Thermodyn.* **2007**, *10*, 1–26.
8. Lozano, M.A.; Valero, A. Theory of the Exergetic Cost. *Energy* **1993**, *18*, 939–960. [CrossRef]
9. Palazzo, P. Thermal and Chemical Aspect in Equation of State and Relation with Generalized Thermodynamic Entropy. *Int. J. Thermodyn.* **2018**, *21*, 55–60. [CrossRef]
10. Moran, M.J.; Sciubba, E. Exergy Analysis: Principles and Practice. *J. Eng. Gas Turbines Power* **1994**, *116*, 285–290. [CrossRef]
11. Barranco-Jimenez, M.A.; Ocampo-Garcia, A.; Angulo-Brown, F. Thermoeconomic optimization of an endoreversible chemical engine model. In *Proc. ECOS 2016*; University of Ljubljana, Faculty of Mechanical Engineering: Ljubljana, Slovenia, 2016; ISBN 9616980157/9789616980159.
12. Barranco-Jimenez, M.A.; Ocampo-Garcia, A.; Angulo-Brown, F. Thermoeconomical analysis of an endoreversible chemical engine model with a diffusive transport law of particles. VIII International Congress of Engineering Physics. *J. Phys. Conf. Ser.* **2017**, *792*, 012097. [CrossRef]
13. Palazzo, P. Hierarchical structure of generalized thermodynamic and informational entropy. *Entropy* **2018**, *20*, 553. [CrossRef]
14. Gaggioli, R.A. Available Energy and Exergy. *Int. J. Appl. Thermodyn.* **1998**, *1*, 1–8.
15. Gaggioli, R.A.; Richardson, D.H.; Bowman, A.J. Available Energy—Part I: Gibbs Revisited. *J. Energy Resour. Technol.* **2002**, *124*, 105–109. [CrossRef]
16. Gaggioli, R.A.; Paulus, D.M., Jr. Available Energy—Part II: Gibbs Extended. *Trans. ASME* **2002**, *124*, 110–115. [CrossRef]
17. Voet, D.; Voet, J.G.; Pratt, G.W. *Fundamentals of Biochemistry: Life at the Molecular Level*, 5th ed.; John Wiley & Sons Inc.: New York, NY, USA, 1999.
18. Demirel, Y. Exergy Use in Bioenergetics. *Int. J. Exergy* **2004**, *1*, 128–146. [CrossRef]
19. Demirel, Y. Nonequilibrium Thermodynamics Modeling of Coupled Biochemical Cycles in Living Cells. *J. Non Newton. Fluid Mech.* **2010**, *165*, 953–972. [CrossRef]
20. Demirel, Y. Thermodynamics Analysis. *Arab. J. Sci. Eng.* **2013**, *38*, 221–249. [CrossRef]
21. Demirel, Y. On the Measurement of Entropy Production il Living Cells under an Alternating Electric Field. *Cell Biol. Int.* **2014**, *38*, 898–899. [CrossRef] [PubMed]
22. Lucia, U.; Grisolia, G. Thermal Resonance and Cell Behavior. *Entropy* **2020**, *22*, 774. [CrossRef]
23. Lucia, U.; Grisolia, G. Non-Equilibrium Thermodynamic Approach to Ca^{2+}-Fluxes in Cancer. *Appl. Sci.* **2020**, *10*, 6737. [CrossRef]
24. Lucia, U.; Grisolia, G. Seebeck-Peltier Transition Approach to Oncogenesis. *Appl. Sci.* **2020**, *10*, 7166. [CrossRef]
25. Lucia, U.; Grisolia, G.; Deisboeck, T.S. Alzheimer's Disease: A Thermodynamic Perspective. *Appl. Sci.* **2020**, *10*, 7562. [CrossRef]

26. Lucia, U. Irreversibility in Biophysical and Biochemical Engineering. *Physica A* **2012**, *391*, 5997–6007. [CrossRef]
27. Lucia, U. The Guy-Stodola Theorem in Bioenergetic Analysis of Living Systems (Irreversibility in Bioenergetics of Living Systems). *Energies* **2014**, *7*, 5717–5739. [CrossRef]
28. Lucia, U. Bioengineering Thermodynamics of Biological Cells. *Theor. Biol. Med. Model* **2015**, *12*, 29–44. [CrossRef] [PubMed]
29. Lucia, U. Bioengineering Thermodynamics: An Engineering Science for Thermodynamics and Biosystems. *Int. J. Thermodyn.* **2015**, *18*, 254–265. [CrossRef]
30. Bejan, A.; Lorente, S. Constructal Theory of Generation of Configuration in Nature and Engineering. *J. Appl. Phys.* **2006**, *100*, 5. [CrossRef]
31. Bejan, A.; Lorente, S. The Constructal Law of Design and Evolution in Nature. *Philos. Trans. R. Soc. B Biol. Sci.* **2010**, *365*, 1335–1347. [CrossRef] [PubMed]
32. Bejan, A.; Lorente, S. Constructal Law of Design and Evolution: Physics, Biology, Technology and Society. *J. Appl. Phys.* **2013**, *133*, 151301. [CrossRef]

Article

Simulating Finite-Time Isothermal Processes with Superconducting Quantum Circuits

Jin-Fu Chen [1,2], **Ying Li** [2] and **Hui Dong** [2,*]

[1] Beijing Computational Science Research Center, Beijing 100193, China; chenjinfu@csrc.ac.cn
[2] Graduate School of China Academy of Engineering Physics, No. 10 Xibeiwang East Road, Haidian District, Beijing 100193, China; yli@gscaep.ac.cn
* Correspondence: hdong@gscaep.ac.cn

Abstract: Finite-time isothermal processes are ubiquitous in quantum-heat-engine cycles, yet complicated due to the coexistence of the changing Hamiltonian and the interaction with the thermal bath. Such complexity prevents classical thermodynamic measurements of a performed work. In this paper, the isothermal process is decomposed into piecewise adiabatic and isochoric processes to measure the performed work as the internal energy change in adiabatic processes. The piecewise control scheme allows the direct simulation of the whole process on a universal quantum computer, which provides a new experimental platform to study quantum thermodynamics. We implement the simulation on ibmqx2 to show the $1/\tau$ scaling of the extra work in finite-time isothermal processes.

Keywords: quantum thermodynamics; quantum circuit; open quantum system; isothermal process; IBM quantum computer

Citation: Chen, J.-F.; Li, Y.; Dong, H. Simulating Finite-Time Isothermal Processes with Superconducting Quantum Circuits. *Entropy* **2021**, *23*, 353. https://doi.org/10.3390/e23030353

Academic Editor: Michel Feidt

Received: 9 February 2021
Accepted: 12 March 2021
Published: 16 March 2021

Publisher's Note: MDPI stays neutral with regard to jurisdictional claims in published maps and institutional affiliations.

1. Introduction

Quantum thermodynamics, originally considered an extension of classical thermodynamics, has sharpened our understanding of the fundamental aspects of thermodynamics [1–6]. Along with the theoretical progress, experimental tests and validations of the principles are relevant in the realm. Simulation of the quantum thermodynamic phenomena [7–10], as one of the experimental efforts, has been intensively explored with specific systems, e.g., a single trapped particle for testing the Jarzynski equation [11,12], the trapped interacting Fermi gas for quantum work extraction [13,14], and the superconducting qubits for the shortcuts to adiabaticity [15,16]. These specific systems often have limited functions to test generic quantum thermodynamic properties. In quantum thermodynamics, the concerned system, as an open quantum system, generally evolves with the coupling to the environment. Simulations of open quantum systems have been proposed theoretically in terms of quantum channels [17–21], and realized experimentally on various systems, e.g., trapped ions [22], photons [23], nuclear spins [24], superconducting qubits [25], and IBM quantum computer recently[26,27]. The previous works mainly focus on simulating fixed open quantum systems, where the parameters of the systems are fixed with the evolution governed by a time-independent master equation. To devise a quantum heat engine, it is necessary to realize tuned open quantum systems to formulate finite-time isothermal processes.

Simulation with generic quantum computing systems shall offer a universal system to demonstrate essential quantum thermodynamic phenomena. Yet, simulation of a tuned open quantum system remains a challenge mainly due to the inability to physically tune the control parameters and the difficulty to measure the work extraction. In quantum thermodynamics, the work extraction, as a fundamental quantity [28–30], requires the tuning of the control parameters. Such requirement is achievable in the specifically designed system, e.g., the laser-induced force on the trapped ion [11], the trapped frequency of the Fermi gas [13,14], and the external field in the superconducting system [15,16]. However, on a universal quantum computer, e.g., IBM quantum computer (ibmqx2), the user is forbidden to tune the actual physical parameters since the parameters have been optimized to reduce errors. An additional problem is the measurement of the work extraction. In classical thermodynamics, it is obtained by recording the control parameters and measuring the conjugate quantities, but such measurement is not suitable in the quantum region [31].

In this paper, an experimental proposal is given to overcome the difficulty in simulating a finite-time isothermal process. We introduce a virtual way to tune the control parameters, i.e., without physically tuning the parameters. The dynamics are realized by quantum gates encoded the parameter change. As a demonstration, we realize the simulation of a two-level system on ibmqx2 [32] for the isothermal processes, which are fundamental to devise quantum heat engines, yet complicated due to the coexistence of the changing Hamiltonian and the interaction with the thermal bath.

To implement the simulation on a universal quantum computer, we adopt a discrete-step method to approach the quantum isothermal process [33–38]: the isothermal process is divided into series of elementary processes, each consisting of an adiabatic process and an isochoric process. In the adiabatic process, the parameter tuning is performed virtually with the unitary evolution implemented by quantum gates. In the isochoric process, the dissipative evolution is carried out with quantum channel simulation [23,25,39–42] with ancillary qubits, which play the role of the environments[18,21,26]. With this approach, we achieve the simulation of the isothermal process on the generic quantum computing system without physically tuning the control parameters. The piecewise control scheme distinguishes work and heat, which are separately generated and measured as the internal energy change in the two processes. In the current simulation, the energy spacing of the two-level system is tuned with the unchanged ground and excited states. The tuning of the energy spacing is virtually performed via modulating the thermal transition rate in the isochoric process.

In our proposal, the simulation with a universal quantum computer brings clear advantages. First, *the arbitrary change of the control parameters* is archived by the virtual tuning via the simulation of corresponding dynamics, avoiding the difficulty in tuning the actual physical system. In turn, the parameters can be controlled to follow an arbitrary designed function. Second, we can realize *the immediate change of environmental parameters*, such as the temperature. The effect of the bath is reflected through the state of the auxiliary qubits, which can be controlled flexibly with quantum gates.

2. Discrete-Step Method to Quantum Isothermal Process

In quantum thermodynamics, the concerned system generally evolves under the changing Hamiltonian while in contact with a thermal bath. The interplay between quantum work and heat challenges to characterize the quantum thermodynamic cycle on the microscopic level, where the classical method to measure the work via force and distance, is not applicable [31]. For the timescale of the tuning far smaller than the thermal bath response time, the evolution is thermodynamic adiabatic, where the heat exchange with the thermal bath can be neglected, and the internal energy changes due to the performed work through the changing control field. The opposite extreme case with the unchanged control parameters is known as the isochoric process, where the internal energy changes are induced by the heat exchange with the thermal bath. Therefore, work and heat are separated clearly in the adiabatic and isochoric processes, and are obtained directly by measuring the internal energy change.

To simulate the general processes on a universal quantum computer, a piecewise control scheme is necessary to express the continuous non-unitary evolution in terms of quantum channels, where the evolution in each period is constructed by the simulations of open quantum systems [21]. To separate work and heat, we adopt the discrete-step method by dividing the whole process into series of piecewise adiabatic and isochoric processes [33–38]. In Figure 1, the discrete-step method is illustrated with the minimal quantum model, a two-level system with the energy spacing $\omega(t)$ between the ground state

$|g\rangle$ and the excited state $|e\rangle$. Such a two-level system can be physically realized with a qubit, as an elementary unit of the quantum computer. For the clarity of the later discussion, we use the term "two-level system" to denote the simulated system and "qubit" as the simulation system hereafter without specific mention.

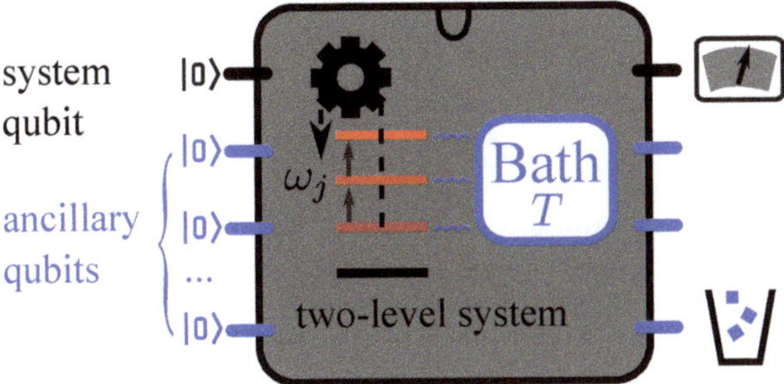

Figure 1. Simulation of the isothermal process on the superconducting quantum computer. The finite-time isothermal process is divided into series of piecewise adiabatic and isochoric processes. In the adiabatic process, the energy of the two-level system is tuned with the switched-off interaction between the system and the thermal bath. In the isochoric process, the interaction is switched on with the unchanged energy spacing ω_j. One qubit represents the simulated two-level system, and the ancillary qubits play the role of the thermal bath at the temperature T. After implementing the quantum circuit, the system qubit is measured to obtain the internal energy.

The state of the two-level system is represented by the density matrix $\rho_s(t)$ of the system qubit, and the thermal bath is simulated by ancillary qubits. Initially, the system qubit is prepared to the thermal state $\rho_s(0)$ at the temperature T. The evolution of the tuned open quantum system is implemented with single-qubit and two-qubit quantum gates. The internal energy of the two-level system is $E(t) = \omega(t)p_e(t)$, where the energy of the ground state is assumed as zero, and the population in the ground (excited) state is $p_g(t) = \langle g|\rho_s(t)|g\rangle$ $(p_e(t) = \langle e|\rho_s(t)|e\rangle)$.

For the system to be simulated, we use the discrete-step method to approach the finite-time isothermal process for the two-level system. The discrete isothermal process contains N steps of elementary processes with the total operation time $\tau + \tau_{\text{adi}}$, where τ (τ_{adi}) denotes the operation time in the isochoric (adiabatic) process. Each elementary process is composed of an adiabatic and an isochoric processes. We set the equal operation time for every elementary process $\delta\tau = (\tau + \tau_{\text{adi}})/N$, with the duration τ/N (τ_{adi}/N) for each isochoric (adiabatic) process.

In the adiabatic process, the system is isolated from the thermal bath and evolves under the time-dependent Hamiltonian. Such a process is described by a unitary evolution with the time τ_{adi}/N. The performed work is determined by the change of the internal energy at the initial and the final time. For a generic adiabatic process, the unitary evolution of the system can be simulated with the single-qubit gate acted on the system qubit. In this paper, we consider the adiabatic process as the quench with zero time $\tau_{\text{adi}} = 0$, occurred at time $t_{j-1} = (j-1)\delta\tau$, $j = 1, 2, ..., N$. As the result of the quench, the energy spacing is shifted from ω_{j-1} to ω_j, while the density matrix $\rho_s(t_{j-1})$ remains unchanged after the quench. At the initial time $t_0 = 0$, the energy is quenched from ω_0 to ω_1 after the initial preparation. The performed work for the quench at time t_{j-1} reads

$$W_j = (\omega_j - \omega_{j-1})p_e(t_{j-1}). \tag{1}$$

To obtain the performed work, we only need to measure the excited state population $p_e(t_{j-1})$ of the system qubit at the beginning of each isochoric process.

In the isochoric process of the j-th elementary process ($t_{j-1} < t \leq t_j$), the two-level system is brought into contact with the thermal bath at the temperature T. The evolution is given by the master equation

$$\dot{\rho}_s = -i[H_j, \rho_s] + \gamma_0 N_j \mathscr{L}(\sigma_+)[\rho_s]$$
$$+ \gamma_0(N_j + 1)\mathscr{L}(\sigma_-)[\rho_s], \tag{2}$$

with

$$\mathscr{L}(\sigma)[\rho_s] = \sigma\rho_s\sigma^\dagger - \frac{1}{2}\sigma^\dagger\sigma\rho_s - \frac{1}{2}\rho_s\sigma^\dagger\sigma. \tag{3}$$

Here, $H_j = \omega_j|e\rangle\langle e|$ is the Hamiltonian of the system during the period $t_{j-1} < t \leq t_j$, $N_j = 1/[\exp(\beta\omega_j) - 1]$ is the average photon number with the inverse temperature $\beta = 1/(k_B T)$, and $\sigma_+ = |e\rangle\langle g|$ ($\sigma_- = |g\rangle\langle e|$) is the transition operator. In this process, the change of the internal energy is induced by the heat exchange with the thermal bath, and no work is performed. During the whole discrete isothermal process, the work is only performed at the time t_j.

We explicitly give the equations for each element of the density matrix according to Equation (2). The populations in the ground and excited states satisfy

$$\dot{p}_g = \gamma_0(N_j + 1)p_e - \gamma_0 N_j p_g, \tag{4}$$

and $p_e = 1 - p_g$. The off-diagonal elements $\rho_{eg}(t) = \langle e|\rho_s(t)|g\rangle$ and $\rho_{ge}(t) = \langle g|\rho_s(t)|e\rangle$ satisfy

$$\dot{\rho}_{eg} = -i\omega_j\rho_{eg} - \gamma_0(2N_j + 1)\rho_{eg}, \tag{5}$$

and $\rho_{ge}(t) = \rho_{eg}^*(t)$. With the unchanged energy eigenstates, the diagonal and the off-diagonal elements of the density matrix evolve separately during the whole isothermal process.

3. Simulation with Quantum Circuits

In this section, we first show the simulation of one elementary process in the circuit. The simulation is formulated for the adiabatic and the isochoric processes as follows.

Adiabatic process. In the superconducting quantum computer, e.g., IBM Q system, the tuning of the physical energy levels of qubits is unavailable for the users. The physical parameters are fixed at the optimal values to possibly reduce noises and errors induced by decoherence and imperfect control.

We consider the Hamiltonian of the simulated two-level system as $H(t) = \omega(t)|e\rangle\langle e|$ with the piecewise tuned energy spacing

$$\omega(t) = \omega_j, \quad t_{j-1} < t \leq t_j \text{ with } j = 1, 2..., N. \tag{6}$$

We will show that the tuning of the energy spacing $\omega(t)$ only affects the thermal transition rate. In the simulation, the thermal transition is simulated through the quantum channel simulation, and can be flexibly modulated by single-qubit gates acted on the ancillary qubits. Therefore, we do not have to physically tune any parameters of the quantum computer, and just algorithmically modulate the simulated thermal transition instead. We propose a virtual tuning of the energy spacing with details explained as follows.

In the virtual process, we need to simulate the unitary evolution of the adiabatic process with single-qubit gates acted on the system qubit. For the adiabatic process, i.e., the quench, the state of the system does not evolve in a short period. We just pretend that the energy of the simulated system is tuned from ω_{j-1} to ω_j in the j-th adiabatic process. This virtual tuning of the energy is reflected by the modulation of the transition rate in the simulation of the isochoric process.

Isochoric process. The dynamical evolution of the isochoric process can be simulated with the generalized amplitude damping channel (GADC)

$$\rho_s(t_j) = \mathscr{E}_{\mathrm{GAD}}^{(j)}[e^{-iH_j\delta\tau}\rho_s(t_{j-1})e^{iH_j\delta\tau}], \tag{7}$$

where $\mathscr{E}_{\mathrm{GAD}}^{(j)} = p_\downarrow^{(j)}\mathscr{E}_\downarrow^{(j)} + p_\uparrow^{(j)}\mathscr{E}_\uparrow^{(j)}$ is divided into two sub-channels, the amplitude damping channel

$$\mathscr{E}_\downarrow^{(j)}[\rho_s] = M_0^{(j)}\rho_s M_0^{(j)\dagger} + M_1^{(j)}\rho_s M_1^{(j)\dagger}, \tag{8}$$

and the amplitude pumping channel

$$\mathscr{E}_\uparrow^{(j)}[\rho_s] = M_2^{(j)}\rho_s M_2^{(j)\dagger} + M_3^{(j)}\rho_s M_3^{(j)\dagger}. \tag{9}$$

The corresponding Kraus operators are $M_0^{(j)} = \cos\theta_j|e\rangle\langle e| + |g\rangle\langle g|$, $M_1^{(j)} = \sigma_-\sin\theta_j$, $M_2^{(j)} = |e\rangle\langle e| + \cos\theta_j|g\rangle\langle g|$ and $M_3^{(j)} = \sigma_+\sin\theta_j$. The coefficient $p_\uparrow^{(j)} = 1/[\exp(\beta\omega_j)+1]$ $(p_\downarrow^{(j)} = 1 - p_\uparrow^{(j)})$ shows the probability of excitation (de-excitation) of the two-level system induced by the thermal bath. The evolution time of the j-th elementary process is encoded in the control parameter θ_j via

$$\cos\theta_j = \exp[-\frac{\gamma_0\delta\tau}{2}\coth(\frac{\beta\omega_j}{2})]. \tag{10}$$

With infinite operation time, the ideal discrete isothermal process is realized by setting $\theta_j = \pi/2$, where the system reaches thermal equilibrium at the end of each isochoric process.

For the initial thermal state $\rho_s(0) = \exp(-\beta H(0))/\mathrm{Tr}[\exp(-\beta H(0))]$, the off-diagonal element remains zero throughout the whole process in the current control scheme. In this situation, the evolution by Equation (7) is simplified as

$$\rho_s(t_j) = \mathscr{E}_{\mathrm{GAD}}^{(j)}[\rho_s(t_{j-1})]. \tag{11}$$

For an initial state with non-zero off-diagonal elements, the off-diagonal elements does not affect the evolution of the populations. This comes from the fact that the diagonal and the off-diagonal elements satisfy separate differential equations by Equations (4) and (5).

Figure 2 shows the quantum circuit to simulate the isochoric process. The two sub-channels $\mathscr{E}_\downarrow^{(j)}$ and $\mathscr{E}_\uparrow^{(j)}$ are realized with an ancillary qubit initially prepared in the ground state. The circuits for these two sub-channels are illustrated in Figure 2a. The meaning of each gate is explained at the bottom of Figure 2. Such simulation circuits are extensively studied in the field of quantum computing and quantum information that we will not explain the setup in detail [40].

To achieve the random selection of the two sub-channels, we design two simulation methods, the hybrid simulation, and the fully quantum simulation, as shown in Figure 2b,c, respectively. The former uses one ancillary qubit for each elementary process under the assist of a classical random number generator (CRNG). The latter utilizes fully quantum circuits with two ancillary qubits for each elementary process. In Table 1, we summarize the simulation procedure for the adiabatic and the isochoric processes.

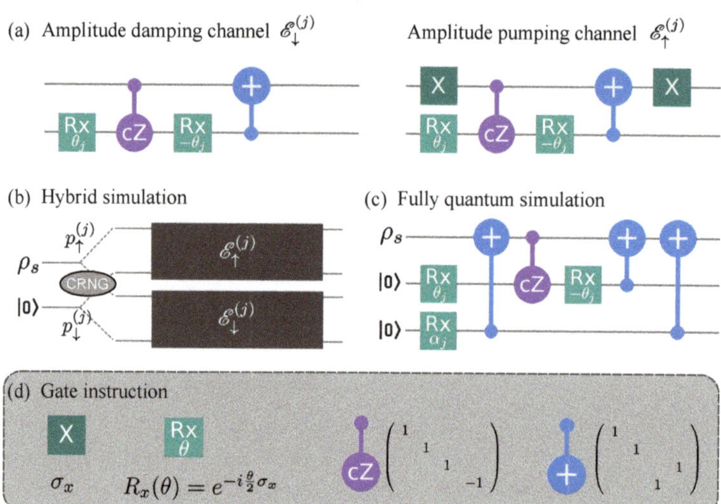

Figure 2. The quantum circuits in one elementary process. (**a**) The amplitude damping (pumping) channel $\mathscr{E}_\downarrow^{(j)}$ ($\mathscr{E}_\uparrow^{(j)}$) in the hybrid simulation. (**b**) One elementary process in the hybrid simulation. The selection of the two sub-channels is realized by the classical random number generator. (**c**) One elementary process in the fully quantum simulation. The selection of the two sub-channels is assisted by another ancillary qubit. (**d**) Instruction of gates in the current simulation.

Table 1. The discrete isothermal process to be simulated and the two simulation methods, the hybrid simulation and the fully quantum simulation

	To be Simulated:	Simulation	
	Discrete Isothermal Process	**Hybrid Simulation with CRNG**	**Fully Quantum Simulation**
Adiabatic process	$U[R(t)], t \in [t_{j-1}, t_j]$	The unitary evolution is realized with the virtual tuning on the system Hamiltonian.	
Isochoric process	System relaxation in Equation (2)	Generalized amplitude damping channel $\mathscr{E}_{GAD}^{(j)}$ with the classical random number generation	Generalized amplitude damping channel $\mathscr{E}_{GAD}^{(j)}$ with an additional qubit at the state $\cos(\alpha_j/2)|0\rangle + i\sin(\alpha_j/2)|1\rangle$
Parameters	Duration: $\delta\tau = t_j - t_{j-1}$ Temperature: T	$\cos\theta_j = \exp[-\frac{\gamma_0\delta\tau}{2}\coth(\frac{\beta\omega_j}{2})]$	$\cos\theta_j = \exp[-\frac{\gamma_0\delta\tau}{2}\coth(\frac{\beta\omega_j}{2})]$ $\cos(\alpha_j/2) = [p_\downarrow^{(j)}]^{1/2}$

3.1. Hybrid Simulation of Isochoric Process with Classical Random Number Generator (CRNG)

With the limited number of qubits, it is desirable to reduce the unnecessary usage of qubits. For the quantum channel of the system qubit, one ancillary qubit is inevitably needed to simulate the non-unitary evolution of the open quantum system [42]. In this method, one qubit represents the two-level system, and each elementary process adds one more ancillary qubit. Therefore, it requires $N + 1$ qubits to simulate the N-step isothermal process.

In the hybrid simulation, the CRNG is used to select the sub-channel $\mathcal{O}_j^{[l]} = \mathscr{E}_\uparrow^{(j)}$ or $\mathscr{E}_\downarrow^{(j)}$ for the isochoric process in the j-th elementary process, as shown in Figure 2b. l denotes the l-th simulation of the discrete isothermal process. For each isochoric process, the CRNG generates a random number $r_j^{[l]} \in [0, 1]$ with uniform distribution. The sub-channel $\mathcal{O}_j^{[l]}$ is selected as $\mathscr{E}_\downarrow^{(j)}$ ($\mathscr{E}_\uparrow^{(j)}$) when the random number satisfies $r_j^{[l]} \leq p_\downarrow^{(j)}$ ($r_j^{[l]} > p_\downarrow^{(j)}$).

3.2. Fully Quantum Simulation of Isochoric Process

For the system with adequate available qubits, the selection of the two sub-channels can be realized on fully quantum circuit by adding two ancillary qubits for each elementary process, as shown in Figure 2c. In each step, one more ancillary qubit is used, prepared to the super-position state $\cos(\alpha_j/2)|0\rangle + i\sin(\alpha_j/2)|1\rangle$ through the $R_x(\alpha_j)$ gate with $\cos(\alpha_j/2) = \sqrt{p_\downarrow^{(j)}}$. This method requires $2N + 1$ qubits to simulate the N-step isothermal process.

Currently, we have solved the problem of separating work and heat. The unitary evolution of the adiabatic process requires isolation from the environment, while the isochoric process needs contact with the environment. Switching on and off the interaction with the thermal bath is complicated and requires enormous efforts, especially in the quantum region for a microscopic system. Fortunately, the design of the quantum computer with a long coherent time ensures the isolation from the environment. The simulation of the quantum channel is designed to simulate the effect of the environment. The advantage of quantum channel simulation over the real coupling to the environment is the flexibility to tune the control parameters, e.g., the temperature, the coupling strength, et al.

The whole evolution of the isothermal process is realized by merging the circuit of each elementary process. In Figure 3, the circuit for the two-step isothermal process is shown as an example. Figure 3a shows the excited state population $p_e(t)$ with the tuned energy spacing $w(t)$ in a two-step isothermal process. The energy spacing is increased from w_0 to w_2 in two steps, while the excited state population decreases from p_0 to p_2.

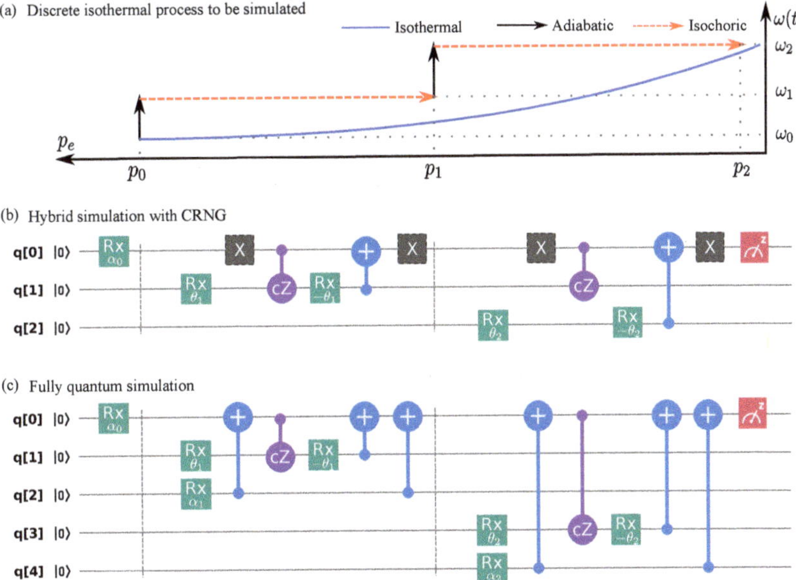

Figure 3. The circuit of the two-step isothermal process on ibmqx2. (**a**) Excited state population-energy ($p_e - E$) diagram. (**b**) The circuit for the hybrid simulation. In each elementary process, the X gate is (or not) implemented for the sub-channel selected as the amplitude pumping (damping) channel according to the classical random number. Each elementary process requires another ancillary qubit. (**c**) The circuit for the fully quantum simulation. Each elementary requires two ancillary qubits.

Figure 3b shows the quantum circuit for the hybrid simulation on ibmqx2. With the five qubits, it is feasible to simulate a four-step isothermal process on ibmqx2. Due to the limited qubit number, the initial state is prepared as a pure state to mimic the thermal state in the current simulation. The populations in the energy eigenstates of the pure state is equal to those of the thermal state, while the non-zero off-diagonal elements lead to the coherence as the superposition of the excited and the ground states. As stated in the description of the isochoric process, such coherence does not affect the evolution of the populations. With another ancillary qubit, a thermal state of the two-level system can be initially prepared through the entanglement between the system and the ancillary qubit.

In the hybrid simulation, the sub-channel $\mathcal{O}_j^{[l]}$ of each elementary process is selected as either the amplitude damping $\mathcal{E}_\downarrow^{(j)}$ or the pumping one $\mathcal{E}_\uparrow^{(j)}$. For an N-step isothermal process, there are 2^N selections of the sub-channels $\{\mathcal{O}_1^{[l]}, \mathcal{O}_2^{[l]}, ..., \mathcal{O}_j^{[l]}, ...\mathcal{O}_N^{[l]}\}$. The circuit of each selection with $N = 2, 3$ and 4 is implemented on ibmqx2. For each selection, the excited state population $p_e^{[l]}(t_j)$ at each step is obtained by repeated implementations of the corresponding circuit. The work in each selection, namely the microscopic work, is given by

$$W^{[l]} = \sum_{j=1}^{N}(\omega_j - \omega_{j-1})p_e^{[l]}(t_{j-1}). \tag{12}$$

The performed work $\overline{W}$ of the whole process is the average of the microscopic work $W^{[l]}$.

Figure 3c shows the fully quantum simulation realized on ibmqx2. With the five qubits, it is possible to realize at most two-step isothermal process, since the qubit resetting process is not permitted on ibmqx2. In the fully quantum simulation, the same circuit is implemented repetitively, and the excited state population $p_e(t_j)$ is obtained by measuring the state of the system qubit. The performed work for the simulated system is given by

$$\overline{W} = \sum_{j=1}^{N}(\omega_j - \omega_{j-1})p_e(t_{j-1}). \tag{13}$$

Since ibmqx2 does not allow the user to reset the state of the qubit, each elementary process requires new ancillary qubit(s). With the ability to reset the ancillary qubit, two (three) qubits are enough to complete the simulation with the hybrid simulation (fully quantum simulation) by resetting the ancillary qubit(s) at the end of each isochoric process. This control scheme is realized in Ref. [25] to simulate repetitive quantum channels on a single qubit.

4. Testing $1/\tau$ Scaling of Extra Work

One possible application of the thermodynamic simulation is to test the $1/\tau$ scaling of the extra work, where τ indicates the operation time of the finite-time isothermal process. In equilibrium thermodynamics, the performed work for a quasi-static isothermal process is equal to the change of the free energy ΔF [43]. The quasi-static isothermal process requires infinite time to ensure equilibrium at every moment. For a real isothermal process, irreversibility arises accompanied with the extra work. For a fixed control scheme, it is proved that the extra work decreases inverse proportional to the operation time at the long-time limit [44–47]. Such $1/\tau$ scaling has been verified for the compression of dry air in the experiment [48].

The superconducting quantum circuit provides an experimental platform to study quantum thermodynamics. We demonstrate the scaling behavior of the extra work in finite-time isothermal process can be observed with the current experimental proposal. Here, the parameters of the simulated two-level system are chosen as $\gamma_0 = 1$ and $\beta = 1$ for convenience. The energy spacing is tuned from $\omega_0 = 1$ to $\omega_N = 2$ in N steps of elementary processes.

In Figure 4, the $1/N$ scaling of the extra work is shown with the ibmqx2 simulation results (Supplementary Materials) for different operation time $\delta\tau = 0.5$ (blue dashed curve) and 10 (red solid curve). For large step number N, it is observed that the extra work is inverse proportional to the step number as $\overline{W} - \Delta F \propto 1/N$ [33,37,38]. The free energy difference of the final and the initial state, namely the performed work in the quasi-static isothermal process is

$$\Delta F = \omega_N - \omega_0 - k_B T \ln \frac{1 + e^{\beta \omega_N}}{1 + e^{\beta \omega_0}}. \tag{14}$$

With the chosen values of the parameters, the explicit value of the free energy difference is $\Delta F = 0.186$. Since the total operation time is $\tau = N \delta \tau$, the $1/N$ scaling is consistent with the $1/\tau$ scaling of the extra work in finite-time isothermal processes. The discrete isothermal processes are simulated on ibmqx2 for $N = 2, 3$ and 4 with the hybrid simulation (empty squares) and $N = 2$ with the fully quantum simulation (pentagrams).

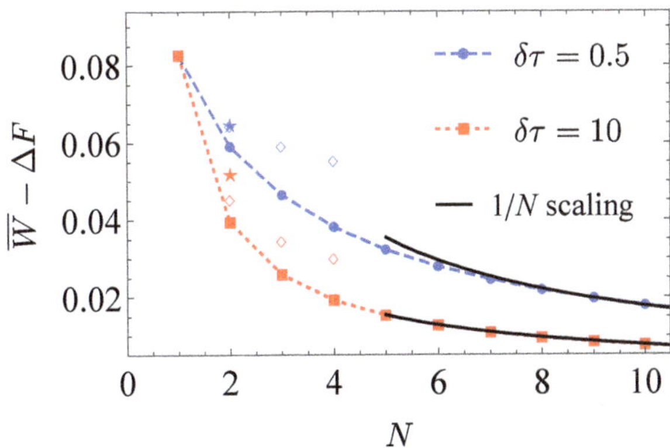

Figure 4. $1/N$ scaling of the extra work for the discrete isothermal process. The operation time of each isochoric process is set as $\delta\tau = 0.5$ (blue dashed curve) or 10 (red solid curve). The ibmqx2 simulation results for $N = 2, 3$ and 4 are plotted. The empty squares present the results by the hybrid simulations, and the pentagrams for the fully quantum simulation. The $1/N$ scaling is shown by the solid black curve.

Figure 5 compares the simulation results on ibmqx2 and the numerical results. In (a) and (b), the work distribution of the hybrid simulation results (blue solid line) is compared to the exact numerical results (gray dashed line), with the operation time $\delta\tau = 0.5$ in (a) and $\delta\tau = 10$ in (b). For the hybrid simulation on ibmqx2, the maximum step number is $N = 4$ with the five qubits. To mimic the random selection of the sub-channel, we simulate every possible selection of the sub-channels in the isochoric processes and measure the state populations of the system qubit. For each selection, the corresponding circuit is implemented on ibmqx2 for 8192 shots. The average work is obtained by summing the work in each selection with the corresponding probability $p_{\{K_j\}} = \prod_j p_{K_j}^{(j)}$ ($K_j = \uparrow$ or $\downarrow$). If the random selections of the sub-channels are possible, $p_{\{K_j\}}$ should be determined by the CRNG. Yet, here the probability of the selection $p_{\{K_j\}}$ is not implemented in the experiment but calculated with $p_{K_j}^{(j)}$ since the random selection of the two sub-channels cannot be implemented on ibmqx2.

Figure 5c,d show the excited state population of the system qubit for the fully quantum simulation of two-step isothermal process on ibmqx2. The operation time of each isochoric process is $\delta\tau = 0.5$ in (c) and $\delta\tau = 10$ in (d). The excited state populations $p_e(t_j)$ at $t_j = 0$, $\delta\tau$ and $2\delta\tau$ are obtained by implementing 40960 shots of the corresponding circuits. Compared to those of the numerical result (gray bar), the ibmqx2 simulation results (blue bar) are larger, since the noises in the quantum computer generally lead to a more mixed state. At the end $t = 2\delta\tau$ of the process, the most quantum gates are used, and the absolute error reaches about 0.05. The fidelity between the simulation and the numerical results $F(t) = \sqrt{p_g^{(num)}(t)p_g^{(sim)}(t)} + \sqrt{p_e^{(num)}(t)p_e^{(sim)}(t)}$ is explicitly $F(2\delta\tau) = 0.998$ and 0.999 for the second step $t = 2\delta\tau$ in (c) and (d), respectively.

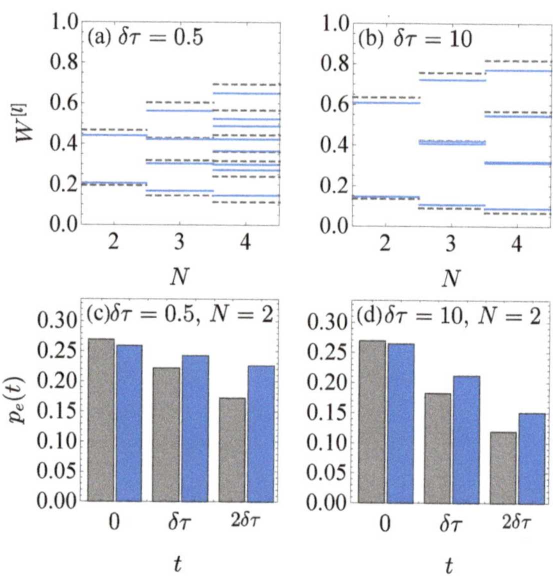

Figure 5. Comparison of the ibmqx2 simulation and the numerical results. (**a,b**) show the microscopic work in the hybrid simulation with the step number $N = 2$, 3 and 4. The ibmqx2 simulation result (blue solid line) is compared with the numerical result (gray dashed line). (**c,d**) show the excited state population $p_e(t)$ at each step in the fully quantum simulation of the two-step isothermal process. The ibmqx2 simulation results (blue bar) are compared to the numerical results (gray bar).

The performed work in both the hybrid simulation and the fully quantum simulation is obtained according to Equations (12) and (13), as listed in Table 2. In Figure 4, the extra work in the ibmqx2 simulation results exceeds that of the numerical result due to the accumulated error in long circuits. The error mainly comes from the two-qubit gates, since the error probability in two-qubit gates (from 1.344×10^{-2} to 1.720×10^{-2}) greatly exceeds that of single-qubit gates (from 3.246×10^{-4} to 2.164×10^{-3}) [32]. The computing accuracy might be improved by using either quantum error correction or quantum mitigation [49]. Limited to the precision of operation on ibmqx2, the results deviate from the theoretical expectations.

Table 2. The performed work obtained by the ibmqx2 simulation and the numerical results.

	N	$\delta\tau = 0.5$		$\delta\tau = 10$	
		$\overline{W}_{\text{ibmqx2}}$	$\overline{W}_{\text{exact}}$	$\overline{W}_{\text{ibmqx2}}$	$\overline{W}_{\text{exact}}$
Hybrid simulation	2	0.251	0.245	0.232	0.226
	3	0.246	0.233	0.221	0.212
	4	0.243	0.224	0.217	0.206
Fully quantum simulation	2	0.251	0.245	0.238	0.226

The current simulation scheme have only considered the commutative Hamiltonian at different steps $[H(t_j), H(t_{j'})] = 0$ and the adiabatic process as the quench with zero time $\delta\tau_{\text{adi}} = 0$. It can also be generalized to the discrete isothermal process with finite-time adiabatic processes, where the effect of the non-commutative Hamiltonian will increase the extra work [50]. For a generic adiabatic process, the unitary evolution of the two-level system should be simulated with the single-qubit gates on the system qubit. The off-diagonal elements of the initial density matrix cannot be neglected, since the changing ground and excited states lead to the interplay between the off-diagonal elements and the populations. Besides, the current simulation can be simplified for the ideal discrete isothermal process, where the perfect thermalization of the isochoric processes allows simulating each elementary process separately by preparing the equilibrium states at the beginning of the adiabatic processes [38].

With the limited number of qubits, we only show a few data points in Figure 4. It requires either more usable qubits or the ability of resetting to simulate the discrete isothermal process with a larger step number N in experiment. Another topic is to test the optimal control scheme [36]. For the given operation time τ, the control scheme is optimized to reach the minimum extra work. The lower bound of the extra work is related to the thermodynamic length [44,46,51,52], which endows a Riemann metric on the control parameter space. The current experimental proposal might also be utilized to measure the thermodynamic length of the isothermal process for the two-level system.

5. Conclusions

We show an experimental proposal to simulate the finite-time isothermal process of the two-level system with the superconducting quantum circuits. Two methods, the hybrid simulation, and the fully quantum simulation, are proposed to realize the generalized amplitude damping channel. Assisted by the classical random number generator or the quantum superposition, the hybrid or the fully quantum simulation can simulate an N-step isothermal process with $N + 1$ or $2N + 1$ qubits, respectively.

We have used the quantum computer of IBM (ibmqx2) to demonstrate the simulation of the discrete isothermal processes, which have been realized for four steps with the hybrid simulation and two steps with the fully quantum simulation. If more steps of elementary processes can be realized experimentally, the $1/\tau$ scaling of the extra work can be tested by the thermodynamic simulation on the universal quantum computer.

Supplementary Materials: The following are available online at https://www.mdpi.com/1099-430 0/23/3/353/s1 for the experimental data on ibmqx2.

Author Contributions: All authors contribute to designing the protocol and writing the paper. J.-F.C. performed the experiments. J.-F.C. and H.D. performed analytical calculations and analyzed data. H.D. conceived the project through discussion with Y.L. All authors have read and agreed to the published version of the manuscript.

Funding: We thank Luyan Sun for helpful discussions at the initial stage of the current work and C. P. Sun for helpful comments. This work is supported by the NSFC (Grants No. 11534002, No.

11875049 and No. 11875050), the NSAF (Grant No. U1930403 and No. U1930402), and the National Basic Research Program of China (Grants No. 2016YFA0301201).

Institutional Review Board Statement: Not applicable.

Informed Consent Statement: Not applicable.

Data Availability Statement: Not applicable.

Conflicts of Interest: The authors declare no competing interest.

References

1. Campisi, M.; Hänggi, P.; Talkner, P. Colloquium: Quantum fluctuation relations: Foundations and applications. *Rev. Mod. Phys.* **2011**, *83*, 771–791. [CrossRef]
2. Esposito, M.; Harbola, U.; Mukamel, S. Nonequilibrium fluctuations, fluctuation theorems, and counting statistics in quantum systems. *Rev. Mod. Phys.* **2009**, *81*, 1665–1702. [CrossRef]
3. Maruyama, K.; Nori, F.; Vedral, V. Colloquium: The physics of Maxwell's demon and information. *Rev. Mod. Phys.* **2009**, *81*, 1–23. [CrossRef]
4. Linden, N.; Popescu, S.; Skrzypczyk, P. How Small Can Thermal Machines Be? The Smallest Possible Refrigerator. *Phys. Rev. Lett.* **2010**, *105*, 130401. [CrossRef] [PubMed]
5. Strasberg, P.; Schaller, G.; Brandes, T.; Esposito, M. Quantum and Information Thermodynamics: A Unifying Framework Based on Repeated Interactions. *Phys. Rev. X* **2017**, *7*, 021003. [CrossRef]
6. Vinjanampathy, S.; Anders, J. Quantum thermodynamics. *Contemp. Phys.* **2016**, *57*, 545–579. [CrossRef]
7. Feynman, R.P. Simulating physics with computers. *Int. J. Theor. Phys.* **1982**, *21*, 467–488. [CrossRef]
8. Lloyd, S. Universal Quantum Simulators. *Science* **1996**, *273*, 1073–1078. [CrossRef] [PubMed]
9. Georgescu, I.M.; Ashhab, S.; Nori, F. Quantum simulation. *Rev. Mod. Phys.* **2014**, *86*, 153–185. [CrossRef]
10. Sandholzer, K.; Murakami, Y.; Görg, F.; Minguzzi, J.; Messer, M.; Desbuquois, R.; Eckstein, M.; Werner, P.; Esslinger, T. Quantum Simulation Meets Nonequilibrium Dynamical Mean-Field Theory: Exploring the Periodically Driven, Strongly Correlated Fermi-Hubbard Model. *Phys. Rev. Lett.* **2019**, *123*, 193602. [CrossRef]
11. An, S.; Zhang, J.N.; Um, M.; Lv, D.; Lu, Y.; Zhang, J.; Yin, Z.Q.; Quan, H.T.; Kim, K. Experimental test of the quantum Jarzynski equality with a trapped-ion system. *Nat. Phys.* **2014**, *11*, 193–199. [CrossRef]
12. Hoang, T.M.; Pan, R.; Ahn, J.; Bang, J.; Quan, H.T.; Li, T. Experimental Test of the Differential Fluctuation Theorem and a Generalized Jarzynski Equality for Arbitrary Initial States. *Phys. Rev. Lett.* **2018**, *120*, 080602. [CrossRef] [PubMed]
13. Deng, S.J.; Diao, P.P.; Yu, Q.L.; Wu, H.B. All-Optical Production of Quantum Degeneracy and Molecular Bose-Einstein Condensation of 6Li. *Chin. Phys. Lett.* **2015**, *32*, 053401. [CrossRef]
14. Deng, S.; Chenu, A.; Diao, P.; Li, F.; Yu, S.; Coulamy, I.; del Campo, A.; Wu, H. Superadiabatic quantum friction suppression in finite-time thermodynamics. *Sci. Adv.* **2018**, *4*, eaar5909. [CrossRef]
15. Zhang, Z.; Wang, T.; Xiang, L.; Jia, Z.; Duan, P.; Cai, W.; Zhan, Z.; Zong, Z.; Wu, J.; Sun, L.; et al. Experimental demonstration of work fluctuations along a shortcut to adiabaticity with a superconducting Xmon qubit. *New J. Phys.* **2018**, *20*, 085001. [CrossRef]
16. Wang, T.; Zhang, Z.; Xiang, L.; Jia, Z.; Duan, P.; Zong, Z.; Sun, Z.; Dong, Z.; Wu, J.; Yin, Y.; et al. Experimental Realization of a Fast Controlled- Z Gate via a Shortcut to Adiabaticity. *Phys. Rev. Appl.* **2019**, *11*, 034030. [CrossRef]
17. Bacon, D.; Childs, A.M.; Chuang, I.L.; Kempe, J.; Leung, D.W.; Zhou, X. Universal simulation of Markovian quantum dynamics. *Phys. Rev. A* **2001**, *64*, 062302. [CrossRef]
18. Wang, H.; Ashhab, S.; Nori, F. Quantum algorithm for simulating the dynamics of an open quantum system. *Phys. Rev. A* **2011**, *83*, 062317. [CrossRef]
19. Sweke, R.; Sinayskiy, I.; Bernard, D.; Petruccione, F. Universal simulation of Markovian open quantum systems. *Phys. Rev. A* **2015**, *91*, 062308. [CrossRef]
20. Shen, C.; Noh, K.; Albert, V.V.; Krastanov, S.; Devoret, M.H.; Schoelkopf, R.J.; Girvin, S.M.; Jiang, L. Quantum channel construction with circuit quantum electrodynamics. *Phys. Rev. B* **2017**, *95*, 134501. [CrossRef]
21. Su, H.Y.; Li, Y. Quantum algorithm for the simulation of open-system dynamics and thermalization. *Phys. Rev. A* **2020**, *101*, 012328. [CrossRef]
22. Schindler, P.; Müller, M.; Nigg, D.; Barreiro, J.T.; Martinez, E.A.; Hennrich, M.; Monz, T.; Diehl, S.; Zoller, P.; Blatt, R. Quantum simulation of dynamical maps with trapped ions. *Nat. Phys.* **2013**, *9*, 361–367. [CrossRef]
23. Lu, H.; Liu, C.; Wang, D.S.; Chen, L.K.; Li, Z.D.; Yao, X.C.; Li, L.; Liu, N.L.; Peng, C.Z.; Sanders, B.C.; et al. Experimental quantum channel simulation. *Phys. Rev. A* **2017**, *95*, 042310. [CrossRef]
24. Xin, T.; Wei, S.J.; Pedernales, J.S.; Solano, E.; Long, G.L. Quantum simulation of quantum channels in nuclear magnetic resonance. *Phys. Rev. A* **2017**, *96*, 062303. [CrossRef]
25. Hu, L.; Mu, X.; Cai, W.; Ma, Y.; Xu, Y.; Wang, H.; Song, Y.; Zou, C.L.; Sun, L. Experimental repetitive quantum channel simulation. *Sci. Bull.* **2018**, *63*, 1551–1557. [CrossRef]
26. Henao, I.; Uzdin, R.; Katz, N. Experimental detection of microscopic environments using thermodynamic observables. *arXiv* **2019**, arXiv:1908.08968.

27. García-Pérez, G.; Rossi, M.A.C.; Maniscalco, S. IBM Q Experience as a versatile experimental testbed for simulating open quantum systems. *NPJ Quantum Inf.* **2020**, *6*, 1. [CrossRef]
28. Alicki, R. The quantum open system as a model of the heat engine. *J. Phys. A Math. Gen.* **1979**, *12*, L103–L107. [CrossRef]
29. Quan, H.T.; Liu, Y.X.; Sun, C.P.; Nori, F. Quantum thermodynamic cycles and quantum heat engines. *Phys. Rev. E* **2007**, *76*, 031105. [CrossRef]
30. Su, S.; Chen, J.; Ma, Y.; Chen, J.; Sun, C. The heat and work of quantum thermodynamic processes with quantum coherence. *Chin. Phys. B* **2018**, *27*, 060502. [CrossRef]
31. Talkner, P.; Hänggi, P. Aspects of quantum work. *Phys. Rev. E* **2016**, *93*, 022131. [CrossRef] [PubMed]
32. IBM Quantum Experience. Available online: https://www.research.ibm.com/ibmq/technology/experience (accessed on 11 March 2021)
33. Nulton, J.; Salamon, P.; Andresen, B.; Anmin, Q. Quasistatic processes as step equilibrations. *J. Chem. Phys.* **1985**, *83*, 334–338. [CrossRef]
34. Quan, H.T.; Yang, S.; Sun, C.P. Microscopic work distribution of small systems in quantum isothermal processes and the minimal work principle. *Phys. Rev. E* **2008**, *78*, 021116. [CrossRef] [PubMed]
35. Anders, J.; Giovannetti, V. Thermodynamics of discrete quantum processes. *New J. Phys.* **2013**, *15*, 033022. [CrossRef]
36. Ma, Y.H.; Xu, D.; Dong, H.; Sun, C.P. Optimal operating protocol to achieve efficiency at maximum power of heat engines. *Phys. Rev. E* **2018**, *98*, 022133. [CrossRef]
37. Bäumer, E.; Perarnau-Llobet, M.; Kammerlander, P.; Wilming, H.; Renner, R. Imperfect Thermalizations Allow for Optimal Thermodynamic Processes. *Quantum* **2019**, *3*, 153. [CrossRef]
38. Scandi, M.; Miller, H.J.D.; Anders, J.; Perarnau-Llobet, M. Quantum work statistics close to equilibrium. *Phys. Rev. Res.* **2020**, *2*, 023377. [CrossRef]
39. Ruskai, M.B.; Szarek, S.; Werner, E. An analysis of completely-positive trace-preserving maps on M2. *Linear Algebra Appl.* **2002**, *347*, 159–187. [CrossRef]
40. Nielsen, M.A.; Chuang, I.L. *Quantum Computation and Quantum Information*; Cambridge University Press: Cambridge, UK, 2009. [CrossRef]
41. Fisher, K.A.G.; Prevedel, R.; Kaltenbaek, R.; Resch, K.J. Optimal linear optical implementation of a single-qubit damping channel. *New J. Phys.* **2012**, *14*, 033016. [CrossRef]
42. Wang, D.S.; Berry, D.W.; de Oliveira, M.C.; Sanders, B.C. Solovay-Kitaev Decomposition Strategy for Single-Qubit Channels. *Phys. Rev. Lett.* **2013**, *111*, 130504. [CrossRef]
43. Callen, H.B. *Thermodynamics And An Introduction To Thermostatistics*, 2nd ed.; Wiley: New York, NY, USA, 1985.
44. Salamon, P.; Berry, R.S. Thermodynamic Length and Dissipated Availability. *Phys. Rev. Lett.* **1983**, *51*, 1127–1130. [CrossRef]
45. Cavina, V.; Mari, A.; Giovannetti, V. Slow Dynamics and Thermodynamics of Open Quantum Systems. *Phys. Rev. Lett.* **2017**, *119*, 050601. [CrossRef] [PubMed]
46. Scandi, M.; Perarnau-Llobet, M. Thermodynamic length in open quantum systems. *Quantum* **2019**, *3*, 197. [CrossRef]
47. Chen, J.F.; Sun, C.P.; Dong, H. Extrapolating the thermodynamic length with finite-time measurements. *arXiv* **2021**, arXiv:2101.02948.
48. Ma, Y.H.; Zhai, R.X.; Chen, J.; Sun, C.P.; Dong, H. Experimental Test of the 1/t Scaling Entropy Generation in Finite-Time Thermodynamics. *Phys. Rev. Lett.* **2020**, *125*, 210601. [CrossRef] [PubMed]
49. Song, C.; Cui, J.; Wang, H.; Hao, J.; Feng, H.; Li, Y. Quantum computation with universal error mitigation on a superconducting quantum processor. *Sci. Adv.* **2019**, *5*, eaaw5686. [CrossRef] [PubMed]
50. Chen, J.F.; Sun, C.P.; Dong, H. Achieve higher efficiency at maximum power with finite-time quantum Otto cycle. *Phys. Rev. E* **2019**, *100*, 062140. [CrossRef]
51. Deffner, S.; Lutz, E. Thermodynamic length for far-from-equilibrium quantum systems. *Phys. Rev. E* **2013**, *87*, 022143. [CrossRef] [PubMed]
52. Crooks, G.E. Measuring Thermodynamic Length. *Phys. Rev. Lett.* **2007**, *99*, 100602. [CrossRef]

MDPI
St. Alban-Anlage 66
4052 Basel
Switzerland
Tel. +41 61 683 77 34
Fax +41 61 302 89 18
www.mdpi.com

Entropy Editorial Office
E-mail: entropy@mdpi.com
www.mdpi.com/journal/entropy